The Anatomy of the Heavens

God's Message in the Stars

The Anatomy of the Heavens

God's Message in the Stars

by John Klein
with Michael Christopher

Published by Covenant Research Institute, Inc.

International Standard Book Number: 978–1-58930-291-4
Library of Congress Control Number: 2013906597

Cover photo by Royce Bair, nightscapephotos.com
Cover design by Katie Klein, www.KatieKleinsmm.com, Bend, Oregon

Printed in the United States of America

For information:
Covenant Research Institute
PO Box 8224
Bend, OR 97708
www.lostintranslation.org

Contents

Constellation Illustrations

Constellation Group

List of Constellations and their Decans

Constellation	Associated Decans
Virgo	Coma, Bootes, Centaurus
Libra	Crux, Victim (Lupus), Corona
Scorpio	Serpens, Ophiuchus, Hercules
Sagittarius	Lyra, Ara, Draco
Capricorn	Sagitta, Aquila, Delphinus
Aquarius	Pisces Austrinus, Pegasus, Cygnus
Pisces	The Band, Andromeda, Cepheus
Aries	Cassiopeia, Cetus, Perseus
Taurus	Orion, Eridanus, Auriga
Gemini	Lepus, Canis Major, Canis Minor
Cancer	Ursa Major, Ursa Minor, Argo
Leo	Hydra, Corvus, Crater

Foreword

This is the fourth book I've worked on as an "editorial partner" with John Klein. The first three dealt with the book of Revelation from an Hebraic perspective, at a detailed, highly illuminating level that I've not seen equaled in any other work by any other author on the same subject. And I've edited several other such books and read many more.

This book is another "first" in the most enlightening terms I can imagine. As always, John Klein comes at his subject from an entirely scientific perspective, which makes loads of sense because his original college degree dealt entirely with science. However, in this case John's approach might be defined as "How God designed the universe to establish, explain, and reinforce His love for, and His ultimate dominion over, the greatest segment of His creation — mankind." And that matchless mixture of science and theology makes this book entirely unique.

Because, even though virtually all of the constellations in the nighttime sky have had fairly well-established "interpretations" associated with each one since ancient days, that's simply not the whole picture. This book is about the "other" tales that clearly predate anything brought forward by mankind to explain what's behind those starry arrangements. As John Klein explains in the following pages, the constellations tell an entirely different, far more credible and far more important story. After all, they were designed by God Himself. And they are integral parts of His overall creation, clearly intended to reinforce everything He desires us to know about Himself and His plans for our redemption.

But please be forewarned! If you're the least bit inclined to suspect that all of this is about astrology, forget it! There's a huge difference between the truths revealed by astronomy and the pseudo-truths associated with any other approach to the way God made the heavens. To put it another way, no one can truly understand any important message if they completely misunderstand — and even denigrate — the originator of that message. Or, much worse, if they sincerely believe that the message came from an entirely different, entirely capricious source.

'Nuff said on the explanatory side of the equation, for John Klein's work needs no explaining at all to anyone willing to simply dive in and read. You'll find it a fascinating piece of work on several levels at once, especially if you're interested in understanding the God of creation in ways you might never have experienced or investigated before.

We already know that He's an awesome God. This book explores an awesome side of Him that most of us have seldom seen. Happy looking!

Michael Christopher
Bend, OR

Introduction

A few years back a biker was riding by a zoo when he saw a little girl leaning too close to the lion's cage. She was trying to get a really good close-up picture with her new camera, but unfortunately she got a bit too close. Suddenly the lion grabbed her by the edge of her jacket and pulled her inside.

This all occurred in full view of her screaming parents and the rest of a large crowd that just happened to be viewing the lions at the same time. But the biker, apparently too energized to feel any fear, jumped off his bike, ran over to the cage, and hit the lion square on the nose with a powerful punch. Whimpering from the pain, the lion let go of the girl and jumped back. The grateful parents could hardly find the words to thank their daughter's savior as he placed their trembling daughter safely into their arms.

Meanwhile, a *New York Times* reporter who had witnessed the whole scene immediately approached the biker and told him, "Sir, that was the most gallant and brave thing I have ever seen a man do in my whole life."

"Why, it was nothing, really! The lion was behind bars. I just saw this little kid in danger and did what I felt was right."

"Well, I'll make sure this won't go unnoticed. I'm a journalist from the *New York Times*, you know, and tomorrow's paper will have this on the front page. What motorcycle do you ride and what affiliations do you have?"

"A Harley Davidson, and I'm a believer in the God of the Bible."

The following morning the biker bought a copy of the *New York Times*. There on the front page, just as promised, was his story with a headline that read:

Radical Christian Biker Gang Member Assaults African Immigrant and Steals His Lunch

In the course of time we have also been shortchanged. Some Christian leaders, guided by our mostly liberal seminaries, have taught a message about the Bible that has focused almost exclusively on the miraculous stories contained in the New Testament. We've also been taught that this testament contains a new covenant that does away with the older covenants and the related teachings found in the older testament. This, they say, has relegated the practice and applicability of the principles of the Old Testament books to a bygone era.

Sadly, these teachings imply that the older books may contain some interesting stories, and we may learn some truths by reading them, but the picture of a personal God portrayed as a caring yet demanding deity — one who loves all of mankind yet still holds each of us accountable for our indiscretions — has been replaced. They describe the newer, kinder, and more personal God of the New Testament as one who loves and forgives mankind first, ahead of all else. Thus they have thrown away the older ideas about God and now teach that we live in an era of endless compassion in which God's limitless grace has almost entirely replaced any kind of reality-based "judgment" at all.

Blinded by Bias

These well-meaning but misguided people have acted just like the journalist in our opening story. Wishful thinking — and the biases it produces — sometimes blind us and utterly preclude a more complete, objective understanding of reality. The comforting blinders they wrap around the eyes of our hearts can lead us into false conclusions about information that might otherwise force us to stop and think a little harder.

However, over the last few decades an important transition has been taking place. Mostly imperceptible at first, this change has begun to pick up momentum. Today, the church is awakening to the reality that maybe "biblical things" are not all as we have been taught. It is now much more common to hear of whole congregations desiring to know more about the biblical feasts, those customs and instructions originally given by God (Yahweh, in Hebrew) in the Old Testament yet mostly ignored until recently. They were observed by Jesus Himself, whom we will call by His actual Hebrew name, *Yeshua*, in the rest of this volume.

What's in a Name?

The world knows Him as "Jesus Christ." But while He lived here on earth He was known by the Hebrew name His mother gave Him at His birth, "Yeshua." As we explained in Volume 1 of the *Lost in Translation* series:

In Luke 1:31, an angel of the Lord told the virgin Mary (actually, in both Hebrew and English her name would be *Miriam* – or *Miryam*, as it's sometimes transliterated) to name her unborn child *Yeshua*, which means "I am Salvation." Mary/Miriam then used that name whenever she called her son in for lunch.

In the 2,000 years since Yeshua's birth, any number of complicated explanations have come along to explain how "Yeshua" became "Jesus Christ." Here's what we consider the most likely sequence:

(Hebrew)		(Greek)		(Latin)		(English)
Yeshua	=	Iesous	=	Jesu	=	Jesus[1]

To the above I would add only one additional point. The "Christ" portion comes from the Greek word "Christos," a title that was often applied to Olympic athletes, meaning "godlike." "Jesus Christ" thus becomes a combination of two separate Latin and Greek words, neither of which Yeshua was ever known by during His life on earth.

Some pastors have even begun to offer Passover Seders to their congregations, copying as closely as they can the ancient practice of celebrating Passover each year. The realization by the modern church that the date of this feast occurred on the very day of our Lord's death on the cross has been one of the catalysts for an amazing renaissance in how we now look at our own Bible. Others have discovered numerous additional insights brought about by reorienting their thinking about the older biblical books. Rather than being replaced by the newer testament — and the "replacement" teachings so many others have advocated for so many years — many modern believers have finally begun to think of the Bible as a *unified compilation of many books* that tells a complete and comprehensive story.

In truth, the complete Bible is a string of separate yet interrelated books given to mankind over the course of many generations. Yet all of those supposedly separate books contribute in their own distinct ways to the unmistakable message God so clearly desires His people to understand and accept. That message begins in the book of Genesis and concludes in the book of Revelation. It contains virtually nothing irrelevant or — worse yet according to some teachers and pastors, "no longer operative" — in any of the books between the beginning and the end.

In essence it's an ongoing revelation from the very mind of God, with an incredible capacity to become more and more relevant, consistent, and rewarding the more we study it. Instead of one set of books replacing an older set, all the books together reveal and describe the workings of God throughout time. And all of it is just as applicable and insightful as when it was first given to man by God.

From this perspective we begin to notice that what God originally started out to do —that is, to offer a means by which fallen man could be redeemed and reunited with Himself — He actually completes in the end. The promises revealed in the New Testament give us hope and are very important. But they are actually nothing more than echoes of very ancient messages, first revealed in the Old Testament (also known as the *Tanakh*).

What God has done is to give us an unfolding plan. We find it first at its very beginning in Genesis, chapters one through three. It takes the form of a simple promise to destroy Satan, the one who deceived Adam and Eve, beginning with the important act of covering mankind's nakedness with the skin of a lamb that had to die because of Adam and Eve's indiscretions. As sad as the death of that innocent lamb was, its effect was only temporary. The connection to a future Lamb who would also have to come — and who would also have to die in order for sin to be permanently done away with — was probably lost on them.

Meanwhile, as time went on and God continued His revelations over the centuries, mankind slowly became aware of the huge consequences of his own actions. There was a price to pay, and only God's own Son would ever be able to pay that price. Most important of all, He would be willing.

One of the reasons the New Testament has been elevated in importance in comparison to the Old Testament is because it tells of the fulfillment of the promise originated by God so many years before. But the coming of the Lamb who would pay the price for our sins is neither the end of the story nor the end of the plan first introduced by God. Not by a long shot.

As we study the biblical narrative we find this plan revealed again and again. But in typical Hebraic fashion, more and more details are added each time the plan comes under discussion. In the final book of the Bible, known to us as Revelation, we finally get a much more complete rendition of His promise to us. Revelation wraps most of the previous biblical prophecies into one compact package. It contains extremely detailed information about how and when different portions of His plan were fulfilled in the past or will be fulfilled in the future. We also learn what part we are to play. Yes, God thinks of us as partners, helpmates, in the fulfillment of our own restoration.

Where Is This Book Going?

The goal of this book will be to focus on God's Master Plan by finding, presenting, and explaining additional information about that plan. But instead of limiting our search to the biblical text, we will expand our view and include messages that God has placed elsewhere in His creation. The first three chapters of the book of Romans suggest that even if mankind did not have God's written Word we would still be without excuse for not recognizing the Creator. We will still be held accountable for knowing and responding to the instructions God reveals to us by what He has enabled us to see with our minds, within our souls, and through our own physical senses. In particular our vision.

If we are going to be held liable for not responding to these messages properly, it would be prudent for us to learn exactly what the instructions and the truths expressed in the creation are. Don't you think?

> [18] For the wrath of God is revealed from heaven against all ungodliness and unrighteousness of men who suppress the truth in unrighteousness, [19] because that which is known about God is evident within them; for God made it evident to them. [20] For since the creation of the world His invisible attributes, His eternal power and divine nature, have been clearly seen, being understood through what has been made, so that they are without excuse.
>
> —Romans 1:18–20

The writer of Romans is suggesting that the reality of God's existence, especially His divine power and loving, generous nature, should be clearly seen by us even if we did not have access to the Bible, or the ability to read it. Thus mankind will be held accountable for the realities that he should be seeing, hearing, and perceiving in other obvious ways as well.

So, what exactly are some of these proclamations? Where are they coming from? And most important of all, how can we gain the perspective to recognize and understand them?

To set the stage, let's review a few biblical principles and then see if we can find some of the information we'll be held accountable for outside of the written word.

Chapter One

What Is God's Plan?

The non-believing world has a great misunderstanding about the church. Unfortunately, much of the fault lies at the feet of the church itself. Most non-Bible-believing people think of the church as a religious body full of hypocrites who practice a form of religion but without any power. They believe that the church teaches its believers that by trying to be good, by practicing religious ceremonies such as attending religious meetings and serving at various religious functions, they can attain a place in heaven in the hereafter. Is it any wonder that non-believers are usually completely ignorant about the true hope and plan that God is actually offering to mankind?

At the same time, many Bible believers are completely ignorant of the existence of a comprehensive plan given to us by God. Or, they understand it only superficially. Unfortunately, being unaware of how God's plan actually works introduces some serious misperceptions into the lives of both unbelievers and uninformed believers as well. As is the case with most plans, whether made by God or man, they usually include responsibilities for all the parties involved.

Certainly our churches teach about the death and resurrection of Christ and the salvation we can receive by believing in Him and accepting

His sacrifice as payment for our sins. But only vaguely does the church enfold this into any *overall* plan as created for us by God. When they leave out *our* part, the doing of which helps complete the plan, our leaders can leave us needlessly weak. That can also make us vulnerable to just about any temptation or deception that might come our way.

Meanwhile, without a proper understanding of the hope founded by God and given to mankind, unbelievers typically perceive the church as nothing more than a group of people practicing *just another pointless form of religion*. They end up seeing the Christian church as pretty much equal to all other forms of religion found in the world today, including those that clearly do not worship the God of the Christian Bible. As a result they completely miss the reality that God's true, overall plan is actually a completely unique idea that's not really "religious" in any derogatory sense whatsoever.

At the very beginning of time God presented His overall plan for His creation. A few days later, Adam and then Eve became aware of what God had in mind. His promises were revealed to them, and they had the opportunity to fully understand the part mankind would play. Mankind's role and responsibilities were made clear.

The Old Testament then became a continuation and further revelation of the details of God's plan. It also records the parts of the plan that unfold within recorded history. The New Testament then continues the same theme by recording historical events that were prophesied earlier, then fulfilled. All of these were included in the original plan as outlined in the Old Testament. The last book of the New Testament, the book of Revelation, concludes the Bible by prophesying about the final events that lead up to the ultimate consummation of God's original plan.

However, the Revelation text is not unique in the sense that it presents entirely new information. On the contrary, it actually complements and completes many other prophetic biblical narratives, and helps to put them in clearer perspective. This is an important point, especially since so few Christians seem to understand it. All of the prophetic events described in the *New* Testament were first revealed in the *Old*

Testament. For example, Genesis 3:6–21 reveals the first part of God's plan for the very first time:

> [6] When the woman saw that the tree was good for food,
> and that it was a delight to the eyes, and that the tree
> was desirable to make one wise, she took from its fruit
> and ate; and she gave also to her husband with her, and
> he ate. [7] Then the eyes of both of them were opened,
> and they knew that they were naked; and they sewed
> fig leaves together and made themselves loin coverings.
> [8] They heard the sound of the LORD God walking in
> the garden in the cool of the day, and the man and his
> wife hid themselves from the presence of the LORD God
> among the trees of the garden.
>
> [9] Then the LORD God called to the man, and said to
> him, "Where are you?" [10] He said, "I heard the sound
> of You in the garden, and I was afraid because I was
> naked; so I hid myself." [11] And He said, "Who told you
> that you were naked? Have you eaten from the tree of
> which I commanded you not to eat?" [12] The man said,
> "The woman whom You gave to be with me, she gave
> me from the tree, and I ate."
>
> [13] Then the LORD God said to the woman, "What is
> this you have done?" And the woman said, "The serpent
> deceived me, and I ate." [14] The LORD God said to the
> serpent, "Because you have done this, cursed are you
> more than all cattle, and more than every beast of the
> field; on your belly you will go, and dust you will eat all
> the days of your life; [15] and I will put enmity between you
> and the woman, and between your seed and her seed;
> He shall bruise you on the head, and you shall bruise
> him on the heel."
>
> [16] To the woman He said, "I will greatly multiply your
> pain in childbirth, in pain you will bring forth children;
> yet your desire will be for your husband, and he will
> rule over you." [17] Then to Adam He said, "Because you
> have listened to the voice of your wife, and have eaten
> from the tree about which I commanded you, saying,

> 'You shall not eat from it'; cursed is the ground because of you; in toil you will eat of it all the days of your life.
> [18] Both thorns and thistles it shall grow for you; and you will eat the plants of the field; [19] by the sweat of your face you will eat bread, till you return to the ground, because from it you were taken; for you are dust, and to dust you shall return."
>
> [20] Now the man called his wife's name Eve, because she was the mother of all the living. [21] The LORD God made garments of skin for Adam and his wife, and clothed them.

Now, please focus your attention on verse 15, which reads: "And I will put enmity between you and the woman, and between your seed and her seed; he shall bruise you on the head, and you shall bruise him on the heel."

Here, God is speaking directly to Satan. He tells Satan that He is going to put a divide between mankind and Satan. The Hebrew word for this divide is *eyva*. *Gesenius*[2] renders this word as "enmity, hostile mind." It comes from the root word *ayav* which means "to be an adversary, to hate, an enemy or foe."

This is exactly what Satan has become. God has given mankind a natural disdain, and sometimes even a natural sense of repulsion, for our adversary. This is even reflected in our natural inclination to fear and pull away whenever we're confronted by a snake, the creature that Satan used to deceive us. Most of us have to work very hard to overcome this natural disdain.

However, if we persist in rejecting the Truth we will slowly succumb to a "new understanding." Unfortunately, this new understanding will be founded on nothing more than a very old proposition from a very old and deceptive being. Unfortunately for many people, when Satan presents himself as the true God, within their minds he literally transforms himself from the snake he is into our supposedly "true creator." Worse yet, we begin to think that we, too, are complete masters of our own lives, able to determine truth for ourselves just as the "god" Satan continually claims to do.

On the Head or the Heel?

The second section of Genesis 3:15 reveals more of God's plan. He tells us that He will bruise Satan on the head, but Satan will bruise God on the heel. *Gesenius* defines the Hebrew word *shoof* which is interpreted by most texts as "bruise." A fuller meaning of this word, however, would be "to lie in wait for anything, to attack, to fall upon anyone." God is telling us that Satan, our deceiver, has a destiny designed and foretold by Himself. Thus He is going to limit this being's time. At the hands of Satan's creator, *Elohim* (the Hebrew word here for God), Satan will be judged and destroyed for what he has done. The symbolic understanding of the head, where Satan receives his wound, adds the meaning that Satan's authority, leadership, and the right to rule or govern will also be destroyed.

This passage reveals the plan and desire of Satan — to rule God's creation. But it also reveals that God has everything under control. All of Satan's and mankind's plans to the contrary will fail. And, it is not difficult to speculate correctly about the adversary's crafty schemes. For example, Revelation reveals that, just before his demise, Satan will proclaim to the entire world that he alone is God. And from Elohim's throne in the temple in Jerusalem he will proclaim his phony "right" to rule all of creation.

This will be the culmination of the counter-covenant plan crafted by Satan. He has lied to himself and to us, saying that we (and of course himself) can become like God. We have the power to rule our own lives, determining what is right and wrong for ourselves. In other words there are no natural laws that are static and come from a higher authority, which we need to respect. We can govern our lives however we want.

Situational ethics and the concept of subjective truths that our modern culture cuddles up to are exactly what Satan sells, and they are what govern his own existence. However, I suspect that the rules of Satan's kingdom, once it is established (however temporarily) over the earth, will not allow the same subjectivity on the part of those who serve him. He will rule with an iron fist. Those who do not fall in line, no matter their situational-ethics perspective or their feel-good beliefs about the nature of truth, will be punished. His kingdom will expand the Muslim punishment of beheading for disobedience.

> Then I saw thrones, and they sat on them, and judgment was given to them. And I saw the souls of those who had been beheaded because of their testimony of Jesus and because of the word of God, and those who had not worshiped the beast or his image, and had not received the mark on their forehead and on their hand; and they came to life and reigned with Christ for a thousand years.
> —REVELATION 20:4

The con is on! Don't be part of it and step into Satan's trap.

Coming to His End

The way in which Satan's kingdom will come to an end is quite clear, but what about all the time, including the past, the present, and what remains of the future, that leads up to these concluding events? He certainly has not been idle for the past six thousand years. He has an unfolding plan with goals along the way. One of his main interim goals involves establishing his footprints in the mind of man.

Getting us to conform to a set of rules different than what the real God has established is extremely important to Satan. But there is still one all-important accomplishment remaining on his to-do list. That is to create the offspring mentioned in Genesis 3:15:

> "And I will put enmity between you and the woman, and between your seed and her seed; He shall bruise you on the head, and you shall bruise him on the heel."

Creating those hybrids is paramount, for he will indwell and give all his power to one of those creatures — as detailed in Revelation 13 (please refer to chapter 2, "The Plan Revealed in the Old Testament," to learn more about these hybrids). This being we know as the False

Messiah, or the Antichrist. He will not appear until sometime in the future, but keeping watch and being ready is the believers' job.

Throughout time, Satan's plan has run parallel with the one originated by God. Opposing and countering God's plan, and attempting to confuse and trump His righteous intentions, have been Satan's goals almost since the dawn of creation. In His Word, God has recorded many of these machinations, which is not only interesting but is also an important piece of evidence for the authenticity of the Bible. Nowhere in recorded history has a people or a nation detailed their own failures, mistakes, defeats, and all the other ugly truths about the abandonment of their principles and their God . . . except for the nation of Israel.

In fact, in ancient recorded history it is very uncommon to find any nation ever changing the identity of the gods worshipped by its own people. Once a group of people begins to worship a group of gods they tend to stay with them whether they ever see any actual benefits or not. Even when such false gods require them to sacrifice their own children, they don't change. Only the collapse of the society itself terminates the worship of these devils masquerading as gods.

However, there is one notable exception, repeated very often. When ever a society began by worshipping the one true God, they usually fell away fairly quickly and become polytheistic, switching their loyalty to many gods. Or at least something other than their first love.

Does Religion Evolve?

Incidentally, it is interesting to note what many evolutionists speculated about years ago as potential proof of the theory of evolution. They believed that as archeologists and researchers went out into the world and studied the religious beliefs of ancient man, they would discover that evolution had taken place there as well.

Their theory went like this. Polytheism would be the most widespread "original" religious structure they would find. Then, as mankind and his societies evolved, so would his religious beliefs. Over time, his original polytheistic beliefs would progress into monotheism because monotheism is certainly a more evolved concept. At least, that's what they believed they would discover.

So what do we know today, after many archeological discoveries over decades of time? The exact opposite has been revealed. As mentioned above, most ancient tribal religions of the world believed in the one true creator God. Yet over time, these ideas devolved into various polytheistic religions. Whether it was the Chinese, various African cultures, or ancient American tribal beliefs, many of these groups began by believing in one creator God. Unfortunately, most of these early beliefs were put aside and replaced with various lies.

Thus millions of people on the earth today, no matter what their heritage might be, now practice religions that offer many gods to choose from. Indeed, some would suggest that Mohammad taught a religion that worships only one god, and at first glance this appears to be correct. But when Mohammad founded Islam he selected the "moon" god from more than 600 alternatives. In fact, when Islam decided to ignore all these other gods, the word "allah" — originally spelled with a lowercase "a" — became "Allah," now usually spelled with a capital "A."

Only the Bible teaches a true monotheistic pathway to God. And only since the death and resurrection of the true Messiah did a countertrend toward monotheism begin.

Multiple Levels of Understanding

Another takeaway from the Genesis 3 passage is that both God and all of mankind have the ability to deliver a head blow to Satan. Yeshua's death on the cross certainly was the focal event. But is there a message here about a portion of God's plan in which we are to play a major part as well?

As my co-author and I have explained in previous books (see *Volume 1, Lost in Translation: Rediscovering the Hebrew Roots of Our Faith*), the biblical Hebrew text works on multiple levels of understanding simultaneously. These layered understandings reveal that we have power and authority too. God's intention is for us to use that power in our daily lives, beginning by opposing sin. We *absolutely can* remove Satan's authority in our lives by ignoring his temptations and deceptions.

However, we have been given a much broader arena to affect once we get our own lives in order. The world around us needs God's standard bearers just as much. How often do we carry His standard for the world to see? Or, have we become experts at blending in, not really revealing the God whom we truly love by hiding His light?

But Satan is going to wound Eve's offspring as well. Who exactly are "Eve's offspring"? The obvious answer is you and I. He tries to wound us every day, by getting us to dwell on — and then to act on — the suggestions he puts in our minds. Granted, it is not a sin to have sinful thoughts come into our minds, but it does become sin when we dwell on or act on them. When we don't oppose Satan's suggestions he is then able to wound us too. This wounding manifests itself in all our affairs, and if not dealt with, it turns into various kinds of bondages in our lives.

At the same time, Yeshua Himself is certainly the most successful "Eve offspring," especially when it comes to confronting sin. He is the most direct meaning of Genesis 3:15, and is the primary target of Satan's relentless attacks. So what do these attacks look like, and have they already occurred?

Satan's Typical Attack

In the above passage the Hebrew word for heel is *akev*. This word means "heel" but also includes the idea of "lie in wait" and "the extreme rear." Thus to give a wound on the heel in this passage conveys the idea of attacking someone from behind, or attacking the rear of an army. These are usually the weakest areas of any armed encampment or troop, for when an army advances it is usually led by its strongest elements. The rear is reserved for the supply units and those responsible for treating the wounded. These are the ones who are least able to resist.

In your personal life, have you ever noticed that temptations usually come when you are tired? Or come most often in areas of temptation that you are weakest in? Think of what happened to Yeshua — right at the very end of 40 days of fasting, when He was at His physically weakest, Satan came and attacked from behind.

He first offered Him bread. How bad could eating some bread be, anyway? Then, knowing that Yeshua's goal was to reclaim authority over His creation, Satan offered Him exactly that . . . but with one condition. Yeshua had to bow down and worship him first. Satan was offering Yeshua what seemed like an easy way to complete the task God had put before Him, but thankfully for our sakes, our Messiah chose not to take the easy way out. That way would not have paid the price that would provide us with the needed covering. Only the payment of blood from the perfect Man would suffice, and Yeshua knew it.

But what is the meaning of the wound on the heel of the Messiah's foot, the one who would come to provide a covering for mankind's sin? Again, Hebrew understanding provides us with some insight. Just as other parts on the body convey meaning, so does the foot. For example, the head conveys the concept of authority; the hand represents the works that one does; the foot reveals a person's path, goal, or destination. Thus Satan's purpose in attacking the heel is to deter God from accomplishing His goal, which was to repair our damaged spirits, souls, and bodies because of the sinful ways mankind has chosen to walk in.

Satan assumed, however, that God's goal was the same as his: to control and rule the world. Yes, both plans did include ruling the creation. But what our adversary wasn't aware of was the primary goal of God's plan — to save mankind from his folly and rebellion by paying the price for his wrongs. In so doing, God would provide the necessary covering.

This covering is a symbol repeatedly used by God throughout the Bible, revealing to mankind what His intentions are. We see it first in Genesis 3 when God used the covering of the lamb skin to hide the results of man's sin. Abraham was told to circumcise himself and his male descendants as a sign of the promised covering that would come. In Egypt, his descendants were told to put a blood covering on the posts of their doors on that first Passover celebration. This again was a reminder to God's people that He would come and save them from their sin, beginning with delivering them from the rebellion of Pharaoh against the admonition of Moses to "Let My people go!"

Finally, He did fulfill all of these prophecies (and promises!) by coming as a man, living a perfect life, and offering Himself as the perfect sacrificial Lamb on the cross. Even so, Satan still attempted to prevent God from accomplishing His goal. He even thought that he had won the battle when he took our Messiah's life on the cross. The goal of God, represented by the heel, appeared to be deterred by Satan when he administered a wound to that heel. What Satan wasn't aware of was God's total and complete commitment to mankind. That came first, even before the real King would come to rule.

We have a God who is different in every possible way from our adversary. Yes, the end result of our salvation, brought about by our Groom, is to restore His creation to its rightful rulers: Himself and mankind. This, by the way, was where we originally started. In the beginning, God gave us the honor and responsibility to share in the management of the earth. But God's primary focus then shifted to giving us back our lives.

> God blessed them; and God said to them, "Be fruitful and multiply, and fill the earth, and subdue it; and rule over the fish of the sea and over the birds of the sky and over every living thing that moves on the earth."
>
> —Genesis 1:28

This job was given to us with the understanding that we would accomplish the task by seeking His wisdom and council, and following through as He suggested. Obviously, we didn't do so well.

How to Complement God's Plan

God's plan was first revealed to mankind almost from the start of our existence. Throughout the ages, as revealed in the biblical text, God gave us additional details about His intentions for His creation and for us. If you have read our other books, especially *Lost in Translation: Rediscovering the Hebrew Roots of Our Faith*, you are aware of the greatest of all gifts given to mankind — restored relationship with God. In the Bible He called this relationship "covenant."

We also learned that there are four relationships to be had through covenant: service, friendship, manager, and bridal. These were given to mankind and represented our part in the restoration process. God has asked us to step into these relationships in a logical sequence, with each one requiring additional responsibilities.

God's plan is complemented when our lives are conformed to His principles. If you haven't read the first book in the *Lost in Translation* trilogy, this might be a good time to do so. It will greatly enhance your insight into what you can do to fulfill His plans for you.

A Special Blessing for Mankind?

Some people believe that God was giving mankind a special blessing in Genesis 1:28:

> God blessed them; and God said to them, "Be fruitful and multiply, and fill the earth, and subdue it; and rule over the fish of the sea and over the birds of the sky and over every living thing that moves on the earth."

They believe that by blessing Adam and Eve God was literally both establishing and endorsing mankind's ability to dominate, control, and help usher righteousness back into the world. Even today, thousands of years later, the same folks believe that mankind is slowly restoring goodness back to the earth through that same God-given blessing.

They see the improvements in technology that enable us to enjoy longer life spans, coupled with other scientific, political, and social advancements, as fulfillments of this blessing. They also believe that, as this process continues, they will be able to help God restore His creation by finally expunging most of the imperfection caused by sin. This will then allow for the completion of God's work, which is to restore the union between God and man.

This, of course, is not my own position. And, as I have stated in other books, it is not God's position either. In Genesis 1:28, the Hebrew word for "be fruitful" is *para*. The primary meaning for this word is "to bear," as in a burden or to bear fruit. Thus Adam and Eve were given a goal to be fruitful, to fill the earth, and to subdue it. Through the righteous management of something someone else created and owned, mankind was to bless the earth.

And yes, it is true that the verse says that God blessed man. However, the Hebrew word translated as "bless" also conveys meanings that do not come through in English. It means "to bend the knee" (to honor someone by bowing down), "to invoke God," "to celebrate" and "to adore."[3]

In other words, God was modeling for Adam how to manage His creation. He was saying to Adam, "I want you to manage my estate though humility and respect. In so doing, fill the earth and subdue it."

The rabbis believe that this is the first commandment given to mankind. In accordance with the divine wish, the world is to be inhabited. And, one who neglects this has abrogated a positive commandment and will incur great punishment, because he thereby demonstrates that he does not wish to comply with the divine will to populate the world.[4]

Yes, God does give us burdens, which have been given to us to carry. And no burden from God is too difficult, for He is a kind and caring God.

Satan attempted to deter God from His goal, represented by the wound on the heel, but in the process he received a deadly head wound. God will step forward and destroy Satan and his plans right at the exact point when the adversary thinks he has finally arrived at the pinnacle of his power. What a great fall this will be, for him as well as his followers.

What Does Science Have to Do with It?

Some people believe that the Bible stands apart from the scientific world. Many times these same people — especially those who simply do not wish to believe in a creator God because of what they believe He might require of them — even suggest that the Bible disagrees with scientific facts.

Actually, many scientific principles were contained first within the biblical text and were validated much later by scientific discoveries, as man's awareness and understanding of his surroundings gradually developed over the centuries. These "biblical-scientific principles" include revolutionary ideas, such as the earth is not flat but is in fact a sphere; the earth orbits around the sun and not the other way around; the universe had a definite beginning and has not always existed. All these — and many more — are supported by the words in the Bible.

The scientific community is aware that there are four forces that manage and control all matter. Whether we consider the movements of planets, the forces involved with the weather on planet earth, or the structure and workings of the atom, these four forces influence all of their interactions. The four forces are gravity, electromagnetism, the weak nuclear force, and the strong nuclear force.

We have also learned that God has given mankind four forces with which to influence his surroundings and complete the work given to him by God through covenant. What God is saying is . . .

1. If you want to be effective in My kingdom, serve others.
2. Making friends is a great way to be influential.
3. Managing other people's concerns and assets well will have many rewards.
4. The marriage relationship is the most intimate, but also the most powerful if you want to influence the world around you.

So . . . the next time you want to accomplish something, start by serving others.

Chapter Two

The Plan Revealed in the Old Testament

God does indeed have a plan. The Old Testament reveals His plan in many ways, by explaining and reinforcing it throughout its pages. Each new revelation seems to add more and more information, which is one of the foundational purposes of the Bible in the first place.

More times than not, pastors and preachers today, when studying and preparing for their lessons for their congregations, will read the New Testament as though it's a stand-alone document. When they run across concepts such as baptism, gifts of the spirit, or prophetic words that seem to describe mysterious images, they use their imaginations and their own cultural perspectives to interpret the text, not realizing that they are ignoring the most obvious foundation for properly interpreting each passage. And that, of course, is the Old Testament.

The key to understanding the New Testament is to realize that all of the writers were Hebrew, and all of them lived in a Hebrew society. And, all of them realized that when they used the word "scripture" they were referring to the Old Testament. These men were using Old Testament verbiage and imagery to convey thoughts so that their eventual readers would be able to research out and discover for themselves that what appears to be brand new concepts were actually very ancient ones first revealed in the Old Testament.

However, the Bible also reveals and confirms its foundational meaning in other ways as well. In our previous books, especially *Lost in Translation, Rediscovering the Hebrew Roots of Our Faith*, we learned that God told Moses to call the first five books of the Bible the *Torah*. I am convinced that, if one wants to be a biblical scholar, New Testament or Old, the key is to begin by fully understanding the Torah, which gives us the foundation for correct interpretation of the entire Bible.

In Hebrew, *Torah* is spelled *tav, vav, reysh, hey*. Each of these letters conveys a phonic sound, as does our English alphabet, but the Hebrew alphabet is also pictographic. This means that each letter in a word conveys an additional meaning that embellishes the overall meaning of each word.

The usual interpretation of the word *Torah* is "the first five books." But the Hebrew meaning is "instruction, law, to be shot through with an arrow, (or to hit the mark)."[5] Hitting the mark, as an archer might do when he shoots arrows at a target, is an excellent picture of the life of Yeshua, our Messiah. He lived a perfect life, hitting the mark by living His life righteously as defined by Torah, the instructions God gave to His people 1,500 years earlier.

However, hidden within this word is a prophecy about Yeshua's life, as well as the plan of God. When you string together the pictographic meanings of the four Hebrew letters that make up *Torah* in Hebrew, the pictures communicate this: "Behold the man nailed to the cross."

Many people scoff when they hear that God created a far-reaching, all-encompassing plan for mankind from the very beginning, and that it has not changed and is still on track. They believe that God has rejected Israel and is now working with the church instead. They also believe that the Old Testament was a failed experiment because the Jews rejected their Messiah and killed Him on a cross. After that, according to the same interpretation, God pursued the Gentiles and formed a new religion around the principles expressed in the New Testament, abolishing the first set of ideas given in the Old Testament.

But this false belief is one of the targets this book will attempt to destroy. Look at the following list of Hebrew names that were given and recorded in Torah. These were the names of the first ten men listed in the biblical text, starting with Adam. All of them would become part of the lineage of the Messiah. They appear in Genesis 5 as well as in Luke 3.

Here are the names:

Diagram 2-1: Names of the Pre-Flood Patriarchs

The Meaning of the Names of the First Ten Pre-Flood Patriarchs			
Patriarch	Meaning of Name	Patriarch	Meaning of Name
Adam	Man	Jared	Shall come down
Seth	Appointed	Enoch	Preaching
Enosh	Mortal	Methuselah	His death shall bring
Kenan	Sorrow	Lamech	The despairing
Mahalale	The blessed God	Noah	Rest

When the meanings of these names are put together in the order in which they appear in Torah, and in which they lived, they communicate that "Man is appointed mortal sorrow, but the blessed God shall come down, preaching that His death shall bring the despairing rest." Even before the Flood, God had a plan. Whether the Jews failed or not (and by the way, all nationalities have failed in one way or another), God is slowly proceeding toward fulfilling the original prophecy buried within the names of these ten patriarchs. Whether we agree with it or do not choose to recognize it, His plan was revealed centuries ago and is still moving inexorably toward completion. The hidden message in these names suggests no disruption, no failure of plan A, and no possibility that it will ever need to be replaced with a different one.

The Audacity of Satan

Even as God works out His own plan, Satan has been busy working out the details of his plan as well. These details are explained by God throughout the biblical text. God does not want us to be ignorant of our adversary's deceptions, so He has exposed Satan's strategies to us even before Satan has "thought them up." Revelation is a perfect example of how God has revealed to mankind what to watch for.

Thus we know what to expect and how to protect ourselves from Satan's wiles. Here God describes end-time events, including Satan's attempt to rule the earth from God's throne from the temple in Jerusalem.

In describing the events that lead up to Satan's rule — and the person who will accomplish this part of Satan's plan — God chose to use what seems to be very difficult language. Some would even call it obscure language. Words describing beasts, wondrous events, and evil entities — such as the woman who is all adorned with jewels in Revelation 17 —often cause the reader to walk away and go read something other than this last book of the Bible.

However, we are learning that the key to understanding these images will not be found in our western mindset, nor in our twenty-first century culture. And certainly not in a study of the Greek language. Hebrew is the key and Torah is the keyhole. The first testament is the foundation; it offers all the cultural, religious, and social perspective the reader needs in order to understand these difficult passages in the newer testament.

God first reveals Satan's plan in Genesis 3, then explains how Satan's offspring attempted to overwhelm the plan of God in Genesis 6. Our adversary's first attempt was defeated by the Flood. This judgment destroyed the Nephilim, Satan's evil offspring, and their attempt to destroy the apple of God's eye and His reason for creating our home, which we call the universe, the Milky Way, and the earth. The Flood saved both mankind and the plans of God. The only plan that seems to have revisions and changes due to its failure is Satan's.

However, Satan has not been deterred. His efforts to recover throughout time have been recorded in the pages of the Tanakh. For example, in the people of Canaan we see the Nephilim charging back into the realm of man right after their first setback. It is not accidental that they are the ones occupying the land God had set aside for Israel, His chosen people. Goliath and King Og, to name just two, are undoubtedly the best-known examples.

But these historical events, as recorded in the Old Testament, do not bring Satan's plan to a close. He will make one final effort that must come to fruition at the end of time. The books of Daniel and Revelation (and others as well) foretell the coming one last time of these evil offspring upon the earth. Here they will once again attempt to steal mankind's identity and authority so they can use these to conquer and rule the earth.

For the seventh but last time, Satan will then establish one last one-world kingdom to suppress, control, and rule the earth. This plan is recorded in the Old as well as in the New Testament. If you want some insight into tomorrow's headlines you can find them previewed in His Word.

Ezekiel 28:13–19, and especially the verses below, contain some very interesting information about Satan, our adversary.

> [13] "You were in Eden, the garden of God; every precious stone was your covering: the ruby, the topaz and the diamond; the beryl, the onyx and the jasper; the lapis lazuli, the turquoise and the emerald; and the gold, the workmanship of your settings and sockets, was in you. On the day that you were created they were prepared. . . .
> [18] By the multitude of your iniquities, in the unrighteousness of your trade you profaned your sanctuaries. Therefore I have brought fire from the midst of you; it has consumed you, and I have turned you to ashes on the earth in the eyes of all who see you."
>
> —Ezekiel 28:13, 18

Satan was created by the hand of God as the most beautiful being of all. But because of Satan's iniquities, God also reveals His destiny – one of eventual consumption by fire. Satan will be destroyed, never again to have a relationship of any kind whatsoever with God or man. In other words, all of Satan's plans will come to naught. His future plans will be cut short just as His earlier efforts were by the Flood God brought upon the earth in ancient times.

God's Truth Revealed

It has been said that one-third of the Tanakh is prophecy. For example, the books of Daniel, Joel, and others lay out God's plans verse-by-verse with alarmingly accurate details.

Daniel, a prophet of God who was exiled to Babylon in the sixth century BC, provides us with an amazing prophecy concerning the two comings of God.

> 25 "So you are to know and discern that from the issuing of a decree to restore and rebuild Jerusalem until Messiah the Prince there will be seven weeks and sixty-two weeks; it will be built again, with plaza and moat, even in times of distress. 26 Then after the sixty-two weeks the Messiah will be cut off and have nothing, and the people of the prince who is to come will destroy the city and the sanctuary. And its end will come with a flood; even to the end there will be war; desolations are determined. 27 And he will make a firm covenant with the many for one week, but in the middle of the week he will put a stop to sacrifice and grain offering; and on the wing of abominations will come one who makes desolate, even until a complete destruction, one that is decreed, is poured out on the one who makes desolate."
>
> —Daniel 9:25–27

In these passages — and especially in Daniel 9 — God reveals the timing of each of these comings. The first was scheduled to occur in 69 weeks, understood by scholars to mean 483 years. However, since these are Hebrew calendar years, 483 years would equal 173,880 days.

In verses 24 through 27, we are informed that these 173,880 days will start with a proclamation to the Jews to return and rebuild Jerusalem. This actually occurred on March 14, 445 BC. Artaxerxes, the king of the Medes and Persians, established a decree that allowed the Jews to return and rebuild Jerusalem. Recall that about 150 years prior,

Nebuchadnezzar, the king of Babylon, conquered Judah and exiled the Jews to Babylon. Over the course of the next 150 years the Medo-Persian Empire conquered the Babylonians, and now Artaxerxes was ruling these lands.

Now, when some scholars add 173,880 days to the date of the proclamation concerning Jerusalem, they reach a very interesting date— the day on which Yeshua made His triumphal entry into Jerusalem, riding on a donkey.[6] This was the day all Jews selected their Passover Lamb, five days before Passover. Here again we see the text from the Old Testament being fulfilled by the events recorded in the New Testament.

On June 28, 1967, another proclamation to return and rebuild Jerusalem was made by the Knesset in Israel. This prophecy introduces a very interesting possibility with respect to the timing of our Lord's second coming. Some interpret this second proclamation to mean that seven weeks, or 49 years, should be added to this 1967 date as well. This 49-year time period would then precede the last week, or the last seven years.

Whether this is true, or whether the last seven years will begin by the signing of a covenant alone, we cannot be sure. Nonetheless, all of these predictions are presented as one original plan, created by God from day one. They have application in the twenty-first century because they did not come to fruition in the first century. There is no implication here that the plan of God was revised or that God had to start over. Rather, the unbiased reader learns of a seamless stream of events that are to unfold throughout time.

All of the Bible's predictions certainly suggest that His second coming is closer than some people believe. But as of this writing it is also later than some others have predicted. For example, God spoke through Isaiah the prophet about 2,600 years ago, revealing events that would come about 600 years hence.

> [6] For a child will be born to us, a son will be given to us; and the government will rest on His shoulders; and His name will be called Wonderful Counselor, Mighty

> God, Eternal Father, Prince of Peace. [7] There will be no end to the increase of His government or of peace, on the throne of David and over His kingdom, to establish it and to uphold it with justice and righteousness from then on and forevermore. The zeal of the LORD of hosts will accomplish this.
>
> —Isaiah 9:6, 7

The boy prophesied above did, in fact, come. He was born of a virgin (Isaiah 7:14), fulfilling another prophetic passage about His first coming. In addition to their historical orientation, other prophecies revealed that He would give up His life on a cross.

However, the death of Yeshua did not terminate the plans of God. Many biblical prophecies concern events that have not yet occurred. They foretold that this boy would grow into a mighty king, ushering upon the earth peace and a kingdom from which He would reign forever. This King will, in fact, be Yahweh/Yeshua, the God of the Old Testament.

The plans of God have been revealed for the benefit of anyone who is willing to listen. Their inspiration and accuracy should be a witness for His existence, inspiring us to seek Him out with all of our energy.

Is God's Plan Conditional?

I propose that if God had two separate plans, one of two situations would exist in the biblical text. The first possibility is that there would be no end-times unfulfilled prophecies in the Old Testament. And yet, in our brief review we have discovered that many exist.

The second possibility is that these Old Testament end-times prophecies would be conditional, based on the Jews' recognition of, and response to, their Messiah. If His people rejected Him, the text would propose that all the end-times Old Testament foreshadowings would be null and void and not part of the overall plan of God anymore. What we find, however, is the exact opposite from Yeshua's own mouth, no less. But more on this in the next chapter.

This two-plan concept, on which modern dispensationalism is based, in which the older plan is replaced by a newer one, is rooted in something other than the biblical text. The plan first revealed in the Old Testament has been, in part, fulfilled in the records of the New Testament. The balance will be brought to completion in our day, the twenty-first century.

Chapter Three

What About the New Testament?

Despite all that we've already pointed out, some would still say that there is nothing in the New Testament to support the idea that the Bible should be viewed as one continuing revelation. Or, if they did agree that the two testaments together make a complete unit, they would argue that the Old Testament is there only to show us what has been replaced. Supposedly, its sole purpose is to reveal how much better God's new plan of redemption is, which revolves around grace instead of the Law.

Is this true? Does the Bible really present two different plans, or dispensations — one that focuses on obedience to a set of standards to live our lives by, and another plan that does away with the standard and provides salvation through grace? I would like to present some simple responses to these questions.

To begin with — and at the risk of repeating myself — the idea that the New Testament presents a second plan that replaces the previous plan presented in the Old Testament is flat-out wrong. Furthermore, it can be easily disproven by the words of Yeshua Himself, as confirmed and supported by the Apostle Paul, the authors of the gospels, and the other New Testament writers.

But before we get to that, what does the New Testament itself say about the Old Testament? Are there any passages that say anything positive, or that suggest that the Old Testament words and ideas are still in effect and applicable to our lives? Was the first plan as presented in the Old Testament ruined by the failure of Israel, so that the Old Testament was then replaced by a new work that focused instead on the Gentiles, giving them a separate opportunity to please God?

What about the Word "Scripture"?

Before we take a look at several passages in the New Testament we need to understand a few points. The word "Scriptures" appears regularly in its text. What is this word referring to? When it is used in the New Testament is it referring to itself? Of course not! The historical books, letters, and epistles did not exist as a composite "work" when these passages were written. They had not been compiled into one book that we now know as the "New Testament." That compilation was not completed until several centuries later.

Instead, the New Testament word "Scriptures" refer to the Tanakh, the Old Testament. So let's find out exactly what is said about the first testament in the second testament.

> You search the Scriptures because you think that in them you have eternal life; it is these that testify about Me.
>
> —John 5:39

Here Jesus is talking to Jews and is commenting on how they sometimes searched the Tanakh to find eternal life. He was correcting them with respect to what would actually provide their salvation. It was, of course, Himself. The Old Testament Scriptures testified and prophesied about Yeshua's coming, His ministry, and His ultimate sacrifice that would save mankind from sin. He was not denigrating the Old Testament. Rather, He was pointing out its purpose. The law was given to the Jews to explain the ways of holy living. Remember James' instructions to be holy? In the passage above, Yeshua is saying that in these same Scriptures you can find salvation, too. Of course, salvation was to be found in the Lamb Himself, as also predicted in the Scriptures.

In fact, more than 400 prophecies in the Tanakh speak about a coming Savior, all of which were fulfilled by the life, death, and resurrection of the Messiah.

> For whatever was written in earlier times was written for our instruction, so that through perseverance and the encouragement of the Scriptures we might have hope.
>
> —ROMANS 15:4

> [14] You, however, continue in the things you have learned and become convinced of, knowing from whom you have learned them, [15] and that from childhood you have known the sacred writings which are able to give you the wisdom that leads to salvation through faith which is in Christ Jesus. [16] All Scripture is inspired by God and profitable for teaching, for reproof, for correction, for training in righteousness; [17] so that the man of God may be adequate, equipped for every good work.
>
> —II TIMOTHY 3: 14–17

These passages, along with many others, propose that there is much to be had by searching out the truths found in the Scriptures/Tanakh. These principles would include instruction in righteousness, correction, and salvation. In fact, the purpose of the Scriptures is to show us how to fulfill all the good works to complete the plan of God for our lives. In these writings we are to find patience, comfort, and hope. Hope for what? Salvation, of course.

So if there are two plans, the first done away with and replaced by a second one, why does the text seem to interlink the concepts of the Old with the New Testament? For example:

> Jesus said to them, "Have you never read in the Scriptures: 'The stone the builders rejected has become the cornerstone; the Lord has done this, and it is marvelous in our eyes'?"
>
> —MATTHEW 21:42, NIV

> But Jesus answered and said to them, "You are mistaken, not understanding the Scriptures nor the power of God."
>
> —MATTHEW 22:29

"How then will the Scriptures be fulfilled, which say that it must happen this way?" . . . [56] "But all this has taken place to fulfill the Scriptures of the prophets." Then all the disciples left Him and fled.

—MATTHEW 26:54, 56

"Every day I was with you in the temple teaching, and you did not seize Me; but this has taken place to fulfill the Scriptures."

—MARK 14:49

They said to one another, "Were not our hearts burning within us while He was speaking to us on the road, while He was explaining the Scriptures to us?"

—LUKE 24:32

And according to Paul's custom, he went to them, and for three Sabbaths reasoned with them from the Scriptures. [11] Now these were more noble-minded than those in Thessalonica, for they received the word with great eagerness, examining the Scriptures daily to see whether these things were so.

—ACTS 17:2, 11

For he powerfully refuted the Jews in public, demonstrating by the Scriptures that Jesus was the Christ.

—ACTS 18:28

But now is manifested, and by the Scriptures of the prophets, according to the commandment of the eternal God, has been made known to all the nations, leading to obedience of faith.

—ROMANS 16:26

For I delivered to you as of first importance what I also received, that Christ died for our sins according to the Scriptures, [4] and that He was buried, and that He was raised on the third day according to the Scriptures.

—1 CORINTHIANS 15:3, 4

We see in these passages that the New Testament says a lot of good things about the Tanakh. The apostles referred to it routinely to give each other encouragement, guidance, and hope for their daily lives. And remember, they were *not* using the word *Scriptures* to refer to each other's writings. The apostles were writing and making these very points after the death and resurrection of Yeshua, making it clear that the Scriptures were valid before Yeshua's life and afterwards too.

What Did Paul Really Mean?

Also, I know that many people believe that Paul taught that the Law/Torah/Old Testament should be abolished, for the death and resurrection of Yeshua supposedly did away with the Law once and for all. Now we live by Grace and are no longer under the Law. But is this really what Paul taught, or should we understand something else from the passages in which some believe Paul was teaching that the Old Testament was no longer relevant?

As I have said before, many well-respected commentators go into great detail showing that this is an incorrect understanding of Paul's teachings. They suggest that all of these "anti-law" passages are taken completely out of context. What Paul is really talking about is the *purpose* of the Law (Tanakh/Torah). The purpose of the Law was never to bring salvation to mankind. Its only purpose was to show us what sin was, and to reveal the way of holiness that mankind is called to walk in.

For those who are still undecided, or who still believe in the two-separate-plan concept of the Scriptures, God has given us a very clear foundation for understanding all of the controversial passages in which Paul's writings have been misread or seem to be contradictory. In Acts 21 Paul is asked exactly what he thinks about the Law and its application to our lives today.

> [20] And when they heard it they began glorifying God; and
> they said to him, "You see, brother, how many thousands
> there are among the Jews of those who have believed, and
> they are all zealous for the Law; [21] and they have been
> told about you, that you are teaching all the Jews who
> are among the Gentiles to forsake Moses, telling them

> not to circumcise their children nor to walk according
> to the customs. [22] What, then, is to be done? They will
> certainly hear that you have come. [23] Therefore do this
> that we tell you. We have four men who are under a vow;
> [24] take them and purify yourself along with them, and pay
> their expenses so that they may shave their heads; and all
> will know that there is nothing to the things which they
> have been told about you, but that you yourself also walk
> orderly, keeping the Law. [25] But concerning the Gentiles
> who have believed, we wrote, having decided that they
> should abstain from meat sacrificed to idols and from
> blood and from what is strangled and from fornication."
> [26] Then Paul took the men, and the next day, purifying
> himself along with them, went into the temple giving
> notice of the completion of the days of purification, until
> the sacrifice was offered for each one of them.
>
> —Acts 21:20–26

In the above passage Paul is being asked to take a Nazarite vow per instructions given in the Old Testament, which is no small thing. This vow included shaving all the hair off his body and abstaining from eating certain very common foods, in this case for three years. Paul went ahead and confirmed his belief in the Law by taking this vow as proof to the people that the reports of his beliefs to the contrary were inaccurate. So, when we read other passages that seem to suggest that Paul disagreed with Torah principles we need to remember what this passage tells us about what Paul truly believed.

With that in mind we simply cannot come to any other conclusion. Paul loved and supported the Law and taught that it was the standard for holy living and *had not expired*. His primary concern with the Law had to do with its purpose. From the very beginning he clearly said that the Law was *not for salvation, but for holiness*.

It has also been proposed that this passage in Acts suggests that Paul didn't really support the Law but didn't want to admit it to those in Jerusalem. Or, that he had some other motive for hiding his true feelings and opinions. Well, most of us would call that a lie. If Paul is misrepresenting his true opinion in this passage, how do we know when he is telling the truth about what he believed and taught? Is he representing his true position here in Acts or, on the contrary, is he telling us what he believes in Romans or Galatians, in passages that seem to contradict what he is saying in Acts?

Keep in mind that almost all of the controversial passages that seem to support the anti-Torah position mentioned above are found in the writings of Paul. On the other hand, what about Yeshua? What did He say about the Law? Matthew 5:17 gives us a very clear answer.

> "Do not think that I came to abolish the Law or the Prophets; I did not come to abolish but to fulfill."
>
> —MATTHEW 5:17

This quote makes clear the purpose of His ministry. Here the Hebrew word for Law is *Torah*. So, Yeshua's purpose for coming is not to abolish the Torah, the Old Testament, but to fulfill it. Remember that the Bible is full of prophecies that Yeshua was fulfilling. From Genesis to Malachi, including the passages relating to the messages of the stars, the seed of the woman versus the snake, and the historical accounts of Israel's exodus from Egypt, we find predictions pointing to the life and purpose of Yeshua.

But let's read on in Matthew.

> 18 "For truly I say to you, until heaven and earth pass away, not the smallest letter or stroke shall pass from the Law until all is accomplished. 19 Whoever then annuls one of the least of these commandments, and teaches others to do the same, shall be called least in the kingdom of heaven; but whoever keeps and teaches them, he shall be called great in the kingdom of heaven."
>
> —MATTHEW 5:18, 19

Here God is informing us of exactly when the Law/Torah will be accomplished. He tells us that it will not be accomplished until heaven and earth pass away. Then, in Revelation 21:1 the Bible tells us that a new heaven and a new earth will be created for us to dwell in after the thousand-year reign, and that the old earth will pass away.

Since this has not yet happened we can assume that the Law/Torah still has standing and applicability in our lives. And this is the opinion of God Himself! He adds an exclamation point by telling us that not even a jot or tittle shall pass from the Law. The jot is the smallest letter, the *yod* in the Hebrew aleph-bet, while the tittle is a decorative mark found above some of the letters in the original Hebrew script.

Thus God is saying that everything, including every little detail in His text, makes a valid point. It is one giant matrix, a knit-together plan that will all be worked out on planet earth before the end of the thousand-year reign.

To Abolish or Fulfill?

Most people seem to agree that the above passage in Matthew is actually saying that God didn't come to abolish the Law, but to fulfill it. However, many people claim that in the "fulfilling" of the Law He "completed" it, thereby making it null and void. This idea proposes that what Yeshua is really saying is, "I didn't come to abolish the Law, but to abolish the Law!" Which, of course, makes no sense at all.

The Greek word for fulfill is *plaroo*. This word means to "make full, to fill up." It is used in the Septuagint for the Hebrew word *mala*. This word complements its Greek counterpart because it also means "to fill, to make full." But it does not mean "to terminate." Therefore, Yeshua is saying here that the goal of His life on earth was (and still is) to bring a full and complete understanding of His Word to all of mankind, and to fulfill the prophecies foretelling His first coming.

All of this information was given by God in the Tanakh prior to Yeshua's birth in Bethlehem. Nothing in His words in Matthew 5:17 is meant to convey the idea that His life and death on earth, some 2,000 years ago, was meant to "terminate" anything except Satan's authority over us due to our sin.

> God never meant to give mankind two separate plans yielding two separate paths into His eternal presence. As Hebrews 13:8 tells us, "Jesus Christ is the same yesterday and today and forever."

There is no such thing as two plans of God, one a failure and the other a success. There is no such thing as a set of God's principles for one group of people and another for other groups of people. Nor is there a separate standard for one era and another standard for another era. The Law was never given to one race but to *all* races. Israel was a composite of many people groups, as revealed in the descriptions of the people who were led out of Egypt by Moses. Here's just one passage containing some of what Moses explained to those people:

> 1 "Now, O Israel, listen to the statutes and the judgments which I am teaching you to perform, so that you may live and go in and take possession of the land which the LORD, the God of your fathers, is giving you. 2 You shall not add to the word which I am commanding you, nor take away from it, that you may keep the commandments of the LORD your God which I command you. 3 Your eyes have seen what the LORD has done in the case of Baal-peor, for all the men who followed Baal peor, the LORD your God has destroyed them from among you. 4 But you who held fast to the LORD your God are alive today, every one of you.
>
> 5 "See, I have taught you statutes and judgments just as the LORD my God commanded me, that you should do thus in the land where you are entering to possess it. 6 So keep and do them, for that is your wisdom and your understanding in the sight of the peoples who will hear all these statutes and say, 'Surely this great nation is a wise and understanding people.' 7 For what great nation is there that has a god so near to it as is the LORD our God whenever we call on Him? 8 Or what great nation is there that has statutes and judgments as righteous as this whole law which I am setting before you today?"
>
> —DEUTERONOMY 4:1–8

The purpose of Israel, including all the races Israel represented at the time, was to be a witness to all the nations so that knowledge of God — and His principles — would spread throughout the world.

Couldn't be much plainer than that.

Chapter Four

The Plan Revealed in the Seven Feasts

In the United States we celebrate several annual holidays that Americans look forward to each year. Independence Day, celebrated on the 4th of July, comes immediately to mind. All of our national celebrations are accompanied by much fanfare and revelry, and rightly so.

The same types of passionate celebrations also occur during the Super Bowl and other sporting events. Sometimes, when I have time to watch, I am truly amazed at the scenes that flash across the TV screen. People who have painted their bodies in various colors spend the day jumping around, screaming at the top of their lungs, and holding up placards expressing their love for their favorite team.

I usually have several different reactions to this. The first one is: You will never get me to do that! The second is laughter. The third one is a bit more serious. How come no one gets that excited about the celebrations God told believers to observe?

It is unfortunate that many believers are unaware of these celebrations, called *feasts* in the Bible. They are aware of Christmas and Easter, but neither of these is mentioned as holy days, or holidays, that we should celebrate.

Why are we so ignorant about these special biblical days? This is one of the results of being shortchanged by our own church leaders, to which I referred in the introduction of this book. The all-too-common misunderstanding of the complementary roles and purposes of the Old Testament versus the New Testament— especially the idea that the New Testament replaced the Old — has put blinders over our hearts and minds. These blinders are so effective that even when we are instructed in the New Testament to celebrate the feasts, or when the very same text tells us that Yeshua and His apostles celebrated them, we are still unable to see and hear correctly. We persist in our stupor, walking around in ignorance to the ways of God.

So what's so important about these feasts, and why should modern, non-Jewish believers care? First of all, God told His people that they should celebrate them *forever*. He did not say that the celebrations should continue only until the Messiah came, or when some other super-significant event happened. He never said that someday they should not be observed because doing so would suddenly be wrong. He very clearly said that they should be observed forever.

Granted, eternity is a long time, especially toward the end.[7] But shouldn't our attitude be that if it pleases God it should please us to obey Him?

I Am One of God's People

Frankly, it is an honor for me and my family to be identified as among His people. Unfortunately, what many churches of today teach about the biblical feasts is that they are *Jewish religious* celebrations that don't have anything to do with Christian believers.

But is that really true? Where, indeed, did the Jews get the idea to celebrate them? Did they make them up? On the contrary, the biblical feasts did not originate with the Jews. They originated in the mind of God, not the mind of one man or even the collective "mind" of some large portion of mankind. More to the point, they are not "Jewish" at all. They are 100% biblical, as ordained by God for His people.

Is a modern Christian one of God's people or not?

But what do these special days have to do with God's plan to restore the right relationship between Himself and mankind? When one understands the underlying meaning of each one of these feasts, a story begins to unfold. In fact, starting with Passover, which occurs on the 14th day of Nisan (the first month on the Hebrew calendar, which is usually around the month of April on our calendar), God begins to reveal this very plan. He tells in this celebration about the coming Lamb, and what His purpose will be.

The last of these feasts occurs in the month of Tishrei, usually around the months of September or October on our calendar. This feast informs us about the final part of God's plan: His coming back to dwell with mankind forever.

But before we describe these feasts in more detail, let's stop and look more closely at the words God Himself used to describe them. In Hebrew, all words either *are* root words or *derive* from root words. So, when we study Hebrew word meanings it is helpful to find the root word. The meaning of the root word gives us a much better understanding of the words that have evolved from them.

The first one is *hag* which means "a festival." In this case the root word is *hagag,* which means "to dance in a circle, to keep a festival, to leap." A related word, *haga*, means "a place of refuge." All of these words taken together give you a better picture of the meaning for "feast." God is telling us that on His feast days He enjoys the same types of extroverted expressions of love that modern sports fans exhibit. In fact, in the practicing of them we find our refuge, and they give us an identity associated with God. Certainly no one else is practicing them.

The second word interpreted as feast is *mo'ed* and is used in Leviticus 23:2. This verse reads, "Speak unto the children of Israel, and say unto them, concerning the feasts [*mo'ed*] of the Lord . . .". The word *mo'ed* means "a set time, a cycle or year, an assembly, an appointed time, an appointed sign, a signal."[8]

So, God is telling us that He has chosen special times that are important to Him. On these days He has scheduled time for you and Him to meet. For you to not show up would be like making an appointment with your dentist and then forgetting about it. Maybe we have good reasons to frustrate our dentist, but to frustrate God?

These appointed times are the very days that He commanded us to look forward to forever. They are integral parts of His overall plan in which He sent His Son to restore our purity so we could be brought back into a right relationship with God.

Remember the plan?

Of course God thinks it is wonderful when we spend time with Him each day. But these seven appointed times, occurring every year, are special days on which He Himself has made specific appointments with us. I think we should show up.

You Have an Appointment with God

Both of the Hebrew words commonly translated as "feast" imply that God established several fixed times when very important events would occur. Some of these events have already occurred and have established the reliability of His Word. We should be getting ready for the ones that remain, for it takes time to prepare. In fact, in the Bible we learn that those who do not prepare will "run out of oil" like the virgins who were depending on their lamps. Or they will fall asleep and bury their talents.

Here is a list of the seven feasts that the Lord instructed us to observe.

Diagram 4-1: The Seven Feasts of God

Biblical/Hebrew Name	English Name	Time of Observance
Pesach	*Passover*	*Nisan/Aviv 14*
Hag HaMatzah	*Feast of Unleavened Bread*	*Nisan/Aviv 15-21*
Bikkurim	*Firstfruits of Barley Harvest*	*The day after the Sabbath during Hag HaMatzah*
Shavuot	*Feast of Weeks/Pentecost*	*Fifty days from Feast of Firstfruits*
Yom Teruah	*Feast of Trumpets*	*Tishrei 1*
Yom Kippur	*Day of Atonement*	*Tishrei 10*
Sukkot	*Feast of Tabernacles/ Booths*	*Tishrei 15-21*

Now let's go through them one by one. However, we won't go into great detail about how each feast is to be celebrated. Instead, our purpose here is to show which part of the plan of God each feast highlights.

The Four Seasons

Passover will always fall in the spring. So, that means that it will usually fall on or after March 20th in the northern hemisphere. This date on the Gregorian calendar is the spring equinox, the first day of spring and the halfway point in our sky that the sun appears to reach.

Each year on December 21 the sun transverses the lowest course in the southern sky. On June 21, the sun reaches the highest course overhead. These dates are referred to as the winter solstice and the summer solstice respectively. They also denote the beginning of each of these seasons as well. The fall equinox then falls on September 21, which marks the first day of fall.

Because the Hebrew calendar is lunar, the first of the month always starts on the sighting of the first sliver of the new moon as per

God's instructions. In the Gregorian calendar, sometimes there is a need for a leap year. We use a leap day about every four years to keep our calendar in sync with the seasons.

The ancient Hebrews used a leap month. They would add a whole month just before the start of Nisan 1, which was the beginning of their new year. This additional month would be added when they determined that on Nisan 1 they were still more than 14 days from the spring equinox. The ancients were always aware of how many days were left in any of the four seasons and were able to predict the beginning of the next season. Seven leap months occur in about every 19 years.

Pesach (Feast of Passover)

Passover occurs on the 14th day of Nisan, which usually falls in April. The first Passover was celebrated by the Israelites on the evening before they departed Egypt. On that evening each family killed and cooked a lamb that had been selected five days before. This lamb was very special. It could not be just *any* lamb; it had to be perfect, without blemish, the best of the flocks.

During its last five days it had lived as a pet in the family's home, becoming part of the family so that its sacrifice would have increased meaning and would not be seen as irrelevant or immaterial in any way whatsoever. When it was slaughtered its blood was collected and placed on the doorpost of the family's home. The Hebrew letter *tav* was drawn in blood on each side post and on the top, above the door itself. The ancient letter tav was shaped like a cross. So these blood markings resembled the picture we see described in the Gospels at the crucifixion of Yeshua, along with the two convicts.

This first Passover took place in about 1446 BC. It is interesting to note that by drawing a tav, or cross, over their doors to protect them from the angel of death, the Israelites were correctly prophesying the form of death of the Lamb who would come. This prediction occurred many centuries before this particular form of capital punishment was even invented.

It also looked forward to the fulfillment of the ancient prophecy in which Eve's offspring received a wound in His heel. The nail that supposedly pierced Yeshua's feet was actually driven into His heel, not His foot. The Romans knew that the heel was a much more secure location for supporting the full weight of the body, which also meant that the pain would therefore be even greater.

For the ancient Israelites, the underlying message of this feast was that someday another Lamb would come and be slaughtered. But this Lamb would be a perfect man; in fact, He would be God Himself. Thus he would provide a blood covering that would take away our sins, saving us from death and restoring our right standing before our Creator, in essence making us perfect just like Him.

Passover is the feast that opens the doorway into the house of the King. It represents salvation to all those who receive it. The promised covering for our sins that was only figuratively offered to Adam and Eve is now available to us. They received skins to hide the result of their sin. Now, all humanity can receive the payment that takes away our sin, once and for all.

Hag HaMatzah (Feast of Unleavened Bread)

The Passover lamb was slaughtered and prepared in the daytime of the 14th of Nisan. It was eaten that evening, as Nisan 14 became Nisan 15 (Hebrew calendar days start at sunset) beginning the Feast of Unleavened Bread. It was celebrated by the ancient Israelites on the day they left Egypt. God told them to prepare unleavened bread to take with them on their journey out of captivity.

Our responsibilities come into focus as we study the implications of this feast. Even today, in preparation for this feast the Jews remove all of the leavened bread from their homes, and for the next seven days they eat only unleavened bread in accordance with the biblical instruction. In fact, they remove all food items containing any leavening agents from their homes — and therefore from their diets — for this week. Why would they do this? Because they realize that God was using leavened bread metaphorically to represent sin.

In other words, God is saying that once you recognize the work of salvation through the blood of the Lamb, it's your turn. Both new believers and more mature ones are all called to work out their victory by becoming holy. Holiness is defined in Torah as living out the principles we find there.

The Hebrew word for holiness is *kadosh*. Its primary meaning, however, is "to be pure." In multiple places in the Old Testament, as well as in the New, God has asked us to be holy because He is holy. In other words, we are to copy and be like Him.

So, what is purity? In Leviticus 19–21, God defines what holiness — or purity — is, in His mind. In that text He gives us detailed instruction on how to be holy. Unfortunately, in abandoning the Old Testament concepts because we consider them nonapplicable for today, the modern Church has again blinded and prevented us from seeing the true pathway that God intended us to walk.

We cannot pay the price for our sins. Only Yeshua can do that, on the cross. But we can fulfill His admonition to be pure. As we win our personal victories over sinful habits, we remove the leaven from our lives.

There is also a larger meaning here. This is the day on which Yeshua died on the cross and was buried. His burial is a picture of Him taking away the consequences of sin for the world and burying them for all time. He was modeling for us the very thing that we should do. At the cross He took our sins upon Himself, and with His death and burial He paid the price that should have been paid by us. To copy Him we should also die to evil thoughts and actions, by removing them from our lives. That's what Paul was referring to when he said, "I affirm, brethren, by the boasting in you which I have in Christ Jesus our Lord, I die daily" (1 Corinthians 15:31).

Mitzrayim is the Hebrew word for Egypt. Sometimes it is used metaphorically in the Bible to represent the world, sin, and all the other actions and people that oppose God. Thus the message is consistent here. The Israelites left Egypt on the very day of the feast that celebrates leaving behind the leaven in their lives — the worldliness and the sin — and committing themselves to following Him.

Bikkurim (Feast of Firstfruits)

This feast always falls on the first Shabbat after the feast of Passover. Among the rabbis there are some disagreements about its timing. However, we believe that the first century Sadducees had it correctly timed. During the exodus from Egypt, the pillar of fire by night and the cloud by day first appeared in the camp of the Israelites. These unique phenomena represented the presence of God. Their very first appearance occurred on Firstfruits. Yeshua, our Messiah, rose from the dead on this day as well. In fact, He was called the firstfruits from among the dead in I Corinthians 15:20.

This feast also celebrated the beginning of the barley harvest. The best of this harvest was brought before the priest at the temple and given to God in thanks for the harvest He provided. It was recognized as a tithe. Just as they would give the first and best of their barley harvest, the Israelites were modeling for us how we should give back to God from our gain. It should not be whatever is left over after we have paid all of our bills at the end of the month. It should be the first payment we make — the first and the best. Then we should have faith that God will bless what remains — and sometimes even increase it — so we can pay our remaining bills.

In like fashion, Yeshua was the first to rise from the dead. He also was the first and the best, a down payment, if you will, for our own hoped-for redemption and victory over death. In our partnership with Him we are called to be holy because He is holy. As Leviticus 11:44 tells us: "For I am the LORD your God. Consecrate yourselves therefore, and be holy, for I am holy."

The above verse, of course, is only one of several biblical verses that make exactly the same point in basically the same words. After our conversion, men and women are called to covenant with God — and thus to become holy — in order to change their lives and the world around them by their actions, "because it is written, 'YOU SHALL BE HOLY, FOR I AM HOLY'" (1 Peter 1:16).

Shavuot (Feast of Weeks/Pentecost)

This feast occurs fifty days after the beginning of the feast of Firstfruits. In ancient times, on this day the priest would wave leavened bread before God in the temple. This bread was made from the newly harvested wheat, so it represented the firstfruits from the fields. In essence it was the exact opposite of the bread featured in the Feast of Unleavened Bread, in which leavening was excluded because it represented sin. Here, the leavening agent in the dough represents an increase of purity, or holiness. We offer it back to Him to show our love and obedience to Him.

This feast is better known in the Christian world as *Pentecost* and celebrates the coming of the Holy Spirit. In the Christian world it's called Pentecost because of the Greek word *pente*, which means "fifty," because it occurred 50 days after the resurrection. It was not accidental that the Holy Spirit decided to come back on Shavuot. On that day, in the Upper Room, Yeshua's disciples were solemnly celebrating the feast of Shavuot.

But the feast of Shavuot also has earlier biblical roots. Originally it was recognized as the day Moses received the Torah. I believe that someday in the future this will be the day on which our Groom, Yeshua, will come for His bride.

Yom Teruah (Feast of Trumpets)

The Hebrews were called to go to war against those who oppose God, and against the temptations and bondages that existed in their own souls. This feast commemorates the ongoing struggle we have in this world. The enemy stands against us but we are called to war against his influences wherever we find them. This is the day on which the rabbis believe the Messiah will return. At this time, along with those who are aligned with Him, He will come back to earth and go to war against Satan, his fallen angels, and those who have joined themselves with Satan. And by the way — God wins! This will be the fulfillment of the long-prophesied restoration of God's reign over His creation.

Notice the progression of the meanings of these feasts in our personal lives. First, salvation is entered into with an expectation from God that we will remove the sin from our lives. God then provides us with hope by giving us the expectation of eternal life, just as He Himself rose from the dead as a picture of His promise to all believers that they will rise from the dead too.

Shavuot adds to this progression by picturing the bride of Christ, now in possession of many good works. In other words, those who have made themselves holy. This fifth feast, Feast of Trumpets, is a picture of a well-trained army and the war that no longer lies internally but now lies externally. The bride of Christ is someone who is expected to be a witness for Him and to go to war with all things that are in rebellion, and therefore are not holy.

Yom Kippur (Day of Atonement)

This feast occurs ten days after Feast of Trumpets and is the day on which we celebrate having our names written in the Book of Life. The Jews celebrate it by wearing white garments and fasting until evening. This copies the ancient habit of brides who lived during the first century. On their wedding day they would fast, mikveh (baptize), and prepare themselves for their weddings by putting on white garments that represented their purity. So we should also prepare ourselves in anticipation of this event.

In the future, God has promised that there will be a reuniting of Himself with mankind. He has used the concept of marriage to develop this idea. Believers in the God of the Bible can look forward to another wedding someday in the future as foretold in the book of Revelation. This will certainly occur on Yom Kippur.

Sukkot (Feast of Tabernacles/Booths)

This is the seventh and final feast that God instructed His people to celebrate each year. It lasts for seven days, plus one. It is a time of thanksgiving honoring the fruit harvest. With this final harvest complete, the ancient Israelites would honor God by recognizing that He was the source of blessing and direction in their lives. This was represented by the ceremonies of the water libation and the lighting of the huge menorahs in the outer courts of the temple.

The name of this feast comes from the huts the Israelites dwelt in while they were in the wilderness. The original *sukkot* were their homes during this time of testing, but also represented the time that God dwelt with them in their midst. It is believed by many scholars that this feast may have been the actual time of the birth of Yeshua. Considering that four other feasts have fallen on major events during Yeshua's life and death on earth, and also realizing that two more of the feasts will occur on the day of His return and the day of judgment, it would make perfect sense for the day of His birth to coincide with the seventh feast (see text box "Did Yeshua Honor the Feasts?").

Sukkot is the time when we celebrate the Light coming and dwelling with us. This feast alludes to the end of Revelation, in chapter 21, when Elohim comes and begins to dwell forever with mankind in His restored creation.

In Conclusion . . .

Because the seven biblical feasts both encompass and reinforce God's plan for our redemption, it seems very doubtful that they could have been abolished by God, as most in the Christian church of today claim. If the feasts have been abolished, shouldn't the completion of God's plan that they refer to also be abolished?

Could it be possible that God abolished all seven of the feasts even though five of them coincide with five extremely important events in the earthly life of Yeshua? Meanwhile, according to the Bible the other two will coincide with His victory over Satan (Feast of Trumpets, the second coming) and the day of judgment (Yom Kippur).

Did Yeshua Honor the Feasts?

The Bible makes it clear that Yeshua Himself honored the seven feasts ordained by God the Father. He even went to the temple on Hanukkah, the Feast of Dedication (John 10:22), a manmade feast still celebrated by the Jews today, which He also considered worthy of remembrance. But the story does not end there.

Four significant events in the earthly life of Yeshua occurred on four of the seven biblical feast days — His death on *Passover,* His burial on *Unleavened Bread,* His resurrection on *Firstfruits*, and His sending of the Holy Spirit on *Shavuot* (known to modern believers as *Pentecost*).

Every bit of evidence that we have also suggests that He was born in the fall on *Sukkot.* Clearly those well-known shepherds would not have been "abiding in the fields" in the middle of the winter, on the 25th of the modern month of December. Their flocks were always brought down to lower elevations to feed off the land without having to endure the winter temperatures and snow storms of the higher elevations. The biblical text also strongly hints that Sukkot will end the thousand-year reign, ushering in the Great White Throne Judgment. In memory of the concept of the feast of Sukkot, God will then begin His dwelling with man on earth once again.

Two of the seven feasts yet to be fulfilled by Yeshua are the *Feast of Trumpets* and *Yom Kippur.* Speculation about the Feast of Trumpets is especially interesting, because that feast day was known idiomatically in ancient Hebrew as "the day and the hour of which no man knows." This designation came about because the ancients did not have the advantage of modern lunar calendars. Thus the Feast of Trumpets officially began on the first day of the month of *Tishrei*, when two or more witnesses were able to confirm before the Sanhedrin in Jerusalem that they had personally observed the new moon in the evening sky.

Once the beginning of the Feast of Trumpets was officially established, the news was communicated by a series of signal fires. The first one was built on the western wall of Jerusalem and then duplicated on the tops of hills in a series of bonfires that moved from the east to the west, alerting the whole nation of Israel. This was accompanied by trumpet blasts, also called *thunders*, on the ancient ram's horn trumpets known as *shofars*. The shofar blasts officially summoned the head of each family to come immediately to the temple.

In the 24th chapter of Matthew, when His disciples asked when He would be coming back, among other things He said the following: "For as lightning that comes from the east is visible even in the west, so will be the coming of the Son of Man" (Matthew 24:27, NIV). This verse is a direct reference to the signal fires mentioned above, as they moved from one hill to the next to alert the people. Because Jerusalem was in the east, the fires would travel down toward the Mediterranean seacoast to the west.

Yeshua will fulfill two major events on Yom Kippur. This holy day will be both the day of judgment and the day of Yeshua's wedding.

Would God still intend to keep every promise He ever made while, at the same time throwing out virtually all the important signposts that He established to be shared with His people?

Both propositions seem extremely problematic. The first question proposes that God will not fulfill His plan. That is not going to happen. God is not a liar. More important, if the plan has been abolished, we're dead . . . forever.

The second question just sounds contradictory. Why would God abolish the very feasts that He told us to observe forever, even as He would still continue to fulfill the exact promises revealed in the practices of those same feasts? He gave them to us to remind us of the hope that lies before us. This hope was intended to aid us through tough times by helping us keep our eyes on the plan. How often are we reminded in Christian churches about the second coming and how this should give us hope? How many times have we been told that all of this should help us press in and *be the best we can be for God?*

I propose that if we observed the feasts that God Himself gave us, that would be all the encouragement and reminder we might need.

Two More Holidays

In addition to the basic feasts mentioned above, most observant Jews celebrate at least two extra annual holidays. However, both of these might be called "manmade" as opposed to "ordained by God."

The oldest of the two, *Purim*, was established to celebrate the victory of Queen Esther and her uncle Mordecai over their tormentor, Haman, who plotted to kill all the Jews in the Persian kingdom. This is a lesson in how our disobedience can affect us in the long term.

Haman was a descendant of the Amalekites. The Amalekites were brought under judgment by God but were not dealt with as God directed Saul, the King of Israel, to do more than 500 years earlier. Because of Saul's disobedience Haman's ancestors were allowed to live, thus eventually bringing Haman into the picture. However, God used this entire story to emphasize that even when we sin He can still save us from its consequences. In the nick of time God provided a way out from Haman's evil schemes and saved His people.

Hanukkah is an eight-day Jewish festival established to celebrate the rededication of the temple in Jerusalem at the successful end of the Maccabean revolt against the Jews' Syrian/Seleucid oppressors, circa 165 BC. This holiday was established after Antiochus Epiphanies, the Seleucid king, was defeated by the ragtag armies of the Jews.

As soon as they retook their temple in Jerusalem the Jews of that era began to repair and cleanse the house of God. They discovered that they only had enough oil to light the Menorah, their seven-light lampstand, for only one day. The miracle occurred when this bit of oil lasted for eight days, allowing the Hebrews to make a new batch of oil and keep the lights of the Menorah lit during this whole period. This feast is celebrated as a remembrance of our thankfulness for God's blessing and deliverance of our ancestors during their times of trouble. Antiochus was completely defeated and died unexpectedly a few years later.

Chapter Five

Genesis 1—Seasons

Early in this book we learned that the feasts were referred to in the biblical text as *mo'ed*, a Hebrew word that means "a set time, a cycle or year, an assembly, an appointed time, an appointed sign, a signal." Genesis 1:14 reads:

> Then God said, "Let there be lights in the expanse of the heavens to separate the day from the night, and let them be for signs and for seasons and for days, and years."

The word translated as "seasons" in this passage is mo'ed. As you can see (and as we explained earlier), this word does not mean "seasons" as implied by many translators. It absolutely does not convey the idea of the four seasons of the year. It actually means that the lights in the sky are given so we can determine set times, such as the annual cycles, appointed times, or signs. The implication is that, from the very beginning, God gave instructions to mankind about these appointed times, these signals or messages in the form of feasts that foretell His coming to die and pay our price for our sins, and then His coming to set up His kingdom on the earth.

In this passage, we are also told that the lights in the sky are for "signs." The Hebrew word for signs is *ot*, spelled, *aleph, vav, tav*. This word conveys the idea of "a standard, a military ensign[9], especially used for each of the twelve tribes of Israel"[10] as in Numbers 2:2. *Ot* appears again in Genesis 17:1–11:

> [1] Now when Abram was ninety-nine years old, the LORD appeared to Abram and said to him, "I am God Almighty; walk before Me, and be blameless. [2] I will establish My covenant between Me and you, and I will multiply you exceedingly."
>
> [3] Abram fell on his face, and God talked with him, saying, [4] "As for Me, behold, My covenant is with you, and you will be the father of a multitude of nations. [5] No longer shall your name be called Abram, but your name shall be Abraham; for I have made you the father of a multitude of nations. [6] I will make you exceedingly fruitful, and I will make nations of you, and kings will come forth from you. [7] I will establish My covenant between Me and you and your descendants after you throughout their generations for an everlasting covenant, to be God to you and to your descendants after you. [8] I will give to you and to your descendants after you, the land of your sojournings, all the land of Canaan, for an everlasting possession; and I will be their God."
>
> [9] God said further to Abraham, "Now as for you, you shall keep My covenant, you and your descendants after you throughout their generations. [10] This is My covenant, which you shall keep, between Me and you and your descendants after you: every male among you shall be circumcised. [11] And you shall be circumcised in the flesh of your foreskin, and it shall be the sign of the covenant between Me and you."

In the above, Abraham is instructed to circumcise himself as a sign (*ot* in verse 11). God was therefore saying that this sign would confirm the covenant between Himself and Abraham — and, by extension, all of mankind. Here the Hebrew word is being used as a proof, or a token, of a fulfillment of a future event, or prophecy. Therefore, circumcision is a sign of the fulfillment of the promise made by God to Abraham and his offspring to multiply their people and to give them the land of Israel forever.

The twelve *otot* (this is the plural form in Hebrew), or ensigns, on the standards that were carried by each of the twelve tribes of Israel, were given to them by God and represented "signs" that spoke to them of promises and prophecies of things to come. God's people were to keep their eyes on these standards as they marched through the wilderness, so they would be encouraged and invigorated by the constant assurance of specific future promises that had already been given to them by God.

The Sign in the Wilderness

The wilderness journey, starting in Egypt and ending in the promised land of Israel, is a well-known part of ancient history. But it also represents the believer's life. We start out in this world in a fallen state as represented by Egypt. We leave the old, sinful ways behind, trusting in Him to deliver and protect us. The ancient path through the wilderness sometimes resembles the modern believers' walk. We have to learn to trust God even while we suffer through trials.

Yet those same trials help bring us to maturity in our relationships with Him. Our hope is always focused on the promise, as visually represented by the ensign (signs) God has placed before us. These tell us that someday our Messiah will come back as foretold, delivering us from death and allowing us to dwell with Him in the Promised Land.

So the question is, what future events are the stars foretelling in Genesis 1:14, quoted at the beginning of this chapter? To answer this question it might be helpful to see what David had to say about the very same subject. In Psalm 19 he makes a very curious statement:

> [1] The heavens are telling of the glory of God; and their expanse is declaring the work of His hands. [2] Day to day pours forth speech, and night to night reveals knowledge. [3] There is no speech, nor are there words; their voice

> is not heard. [4] Their line has gone out through all the earth, and their utterances to the end of the world. In them He has placed a tent for the sun, [5] which is as a bridegroom coming out of his chamber; it rejoices as a strong man to run his course. [6] Its rising is from one end of the heavens, and its circuit to the other end of them; and there is nothing hidden from its heat. [7] The law of the LORD is perfect, restoring the soul; the testimony of the LORD is sure, making wise the simple.
>
> —Psalm 19:1–7

Apropos of the above verses, most people certainly don't hear any voices. And exactly what knowledge is pouring forth from the heavens? It seems that David is building onto the ideas expressed by God in the first chapter of Genesis. These heavenly bodies are signs referring to prophecies made by God. They are promises to mankind revealing events that will unfold in the future.

Paul, writing in the New Testament, alludes to David's psalm and adds the following:

> [18] For the wrath of God is revealed from heaven against all ungodliness and unrighteousness of men who suppress the truth in unrighteousness, [19] because that which is known about God is evident within them; for God made it evident to them. [20] For since the creation of the world His invisible attributes, His eternal power and divine nature, have been clearly seen, being understood through what has been made, so that they are without excuse. [21] For even though they knew God, they did not honor Him as God or give thanks, but they became futile in their speculations, and their foolish heart was darkened.
>
> —Romans 1:18–21

This passage, in part, is paraphrasing Psalm 19. Here, God emphasizes once again, but this time through David, that the heavens reveal God's existence through what He has made. In fact this passage adds some very interesting statements regarding what was created on the fourth day. It says that the lights up in the sky, including the sun, moon, and the starry host, have voices that speak to us about the knowledge of God.

Now, I don't know about you but usually I don't hear voices when I go out at night and look up. And I think they have special places for people who do. But this is what God says in His Word. So what gives?

Darkening Our Minds

It seems that Paul is suggesting that these signs reveal knowledge about God and His existence. However, many people go to great lengths to blind and darken their minds so they can ignore God's truths. But they do this at their own great peril. This passage says that God is clearly discernible so that mankind will have no excuse for not believing the truth about which these signs speak. In fact, Romans 2:12–15 builds onto this concept even further:

> [12] For all who have sinned without the Law will also perish without the Law, and all who have sinned under the Law will be judged by the Law; [13] for it is not the hearers of the Law who are just before God, but the doers of the Law will be justified. [14] For when Gentiles who do not have the Law do instinctively the things of the Law, these, not having the Law, are a law to themselves, [15] in that they show the work of the Law written in their hearts, their conscience bearing witness and their thoughts alternately accusing or else defending them.

Here it states that God's laws are written in our hearts. Our consciences have been made to know His truths and they will bear witness either for or against us. Furthermore, the evidence revealed by God's handiwork is said to be His Law. The underlying Hebrew word here in Romans 2, for Law, is *Torah*. This same word was also used in Psalm 19:7 as quoted earlier.

The rest of Psalm 19 establishes the same comparison between the Torah and God's creation, especially the stars in the heavens. These biblical texts are confirming that the Torah and the creation — especially the stars — all speak about the very same thing. Together they provide a set of signs that are evident to everyone, telling us about a

God who is not the least bit shy about revealing Himself and His ways to mankind. Indeed, the evidence He provides, on a constant basis, will automatically eliminate any excuses we can come up with for not knowing and obeying Him.

I am certainly aware that God seems to be more obvious to me when I am outdoors. And I am not alone in this experience. Many people can relate. I don't believe, however, that all by itself this is the complete manifestation and fulfillment of these passages. I believe that there are other messages, containing or referring to various promises that God has made to mankind. Thus these starry signs in the sky speak of an ancient prophecy about the plans of God.

It is very possible that these plans could be fulfilled in our lifetimes. If that is so, it would behoove all of us to become as aware of these messages as possible because we will be held accountable for the instructions they reveal.

So — exactly what do the Law and the stars communicate to us about God's plans and purpose? Granted, we have become aware of some of the purposes expressed in the Law about God's plan.

But what are the stars also communicating?

Chapter Six

The Israelite Camp Ensigns

As Israel wandered in the wilderness for 40 years, God gave them very specific instructions about the way their camp was to be organized. It was not just a clump of tents or some haphazard grouping of families. The tabernacle was the place where God dwelt, so it was located in the very center of the camp. This placement teaches us where God wants to be in our lives. He wants to be our first priority, right in the middle.

God then instructed the Israelites to place the tribe of Levi so that they would immediately surround the tabernacle. They had sole responsibility to clean and repair the tabernacle's accoutrements, including the components of the structure itself and all its accessories that were used by the cohenim in the services they performed.

It was the Levites' responsibility to carry all these things wherever the children of Israel traveled. They were to take the tabernacle down and set it up over and over again, as needed, so they could fulfill their even more important function to serve God within the tabernacle.

The tribe of Levi included four families: Merari, Kohath, Gershon, and Moses and Aaron and his sons. The first three were to camp on the north, south, and west sides respectively, around the tabernacle, with Moses and Aaron and his sons camped in front of the gate of the tabernacle, which faced eastward. The tribe of Moses and Aaron eventually became known as the cohenim.

At that point there were still twelve remaining tribes, because the tribe of Joseph had been split into two tribes as represented by his two sons, Manasseh and Ephraim. These twelve tribes were then organized into four subsections headed by four "primary" tribes led by Judah, Reuben, Ephraim, and Dan. Each of these four subsections was assigned two of the remaining eight tribes. God instructed each tribe to camp on a specific side of the tabernacle, three tribes on each of the four sides, around the four families of the centrally placed tribe of Levi.

Thus the Israelites were all arranged around the center of the camp containing the tabernacle and the Levites, with roughly equal numbers on the north, south, west, and east sides. In so arranging themselves they formed "spokes" that pointed out from the center in all four directions of the compass.

It is startling to learn that these twelve tribes were also known by their constellations, which was on the tribal ensign that God instructed each tribe to carry (Numbers 2). Here are the constellations associated with each tribe.

Diagram 6-1: The Camp of Israel

Tribes to the East	
Tribal Name	**Constellation**
Judah	Leo
Issachar	Cancer
Zebulun	Virgo
Tribes to the South	
Tribal Name	**Constellation**
Reuben	Aquarius
Simeon	Pisces
Gad	Aries
Tribes to the West	
Tribal Nameg	**Constellation**
Ephraim and Manasseh	Taurus
Levites	Libra
Benjamin	Gemini
Tribes to the North	
Tribal Name	**Constellation**
Dan	Scorpio
Asher	Sagittarius
Naphtali	Capricorn

These signs are alluded to by Jacob in Genesis 49.

How Many Israelites Were There?

The diagram below summarizes the population totals of the tribes listed above, giving proportional space for the combined populations of each subsection of tribes. As you can see, the camp of the Israelites was shaped like a cross. And, the group headed by the tribe of Judah, which had the largest population, extended outward the farthest and thereby most resembled the longer, lower portion of the cross on which Yeshua was crucified.

Diagram 6-2: The Organization of the Tabernacle[11]

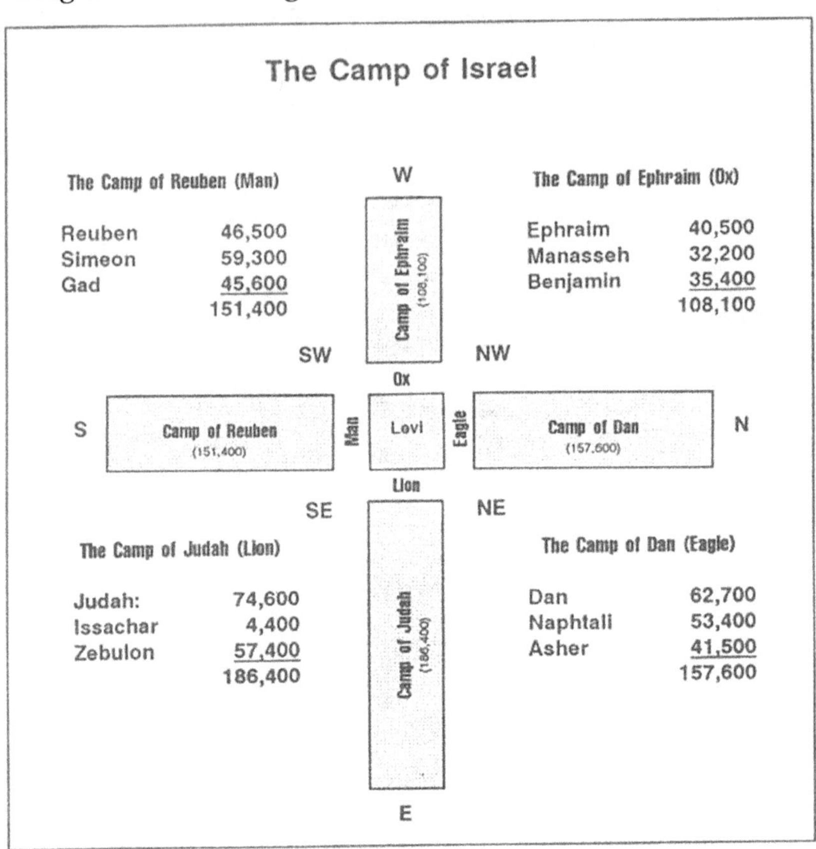

Is it any wonder that Balaam, in Numbers 22–24, was unable to curse the camp? With the fire of God at the center, and the lights of all the campfires of the twelve tribes stretching out to the east, west, north, and south, this would have been a sight to behold.

In fact, even though Balak had promised to reward Balaam for cursing the Israelites, after viewing their camp Balaam chose instead to bless them as God had instructed him. Beyond the awesome number of people involved — and in spite of the demanding logistics that also must have been involved — the Israelite camp was perfectly organized in the shape of a *tav.* This is the Hebrew letter that represents the covenant between God and Israel. Thus Balaam had an additional motive for obeying God and not cursing the Israelites. Smart guy, huh?

Keep in mind that the Old Testament is using imagery (the cross) that some commentators believe has its roots only in the New Testament. But once again this perspective is simply wrong. The cross (or tav) is a very ancient shape and was used to signify the promises and plans of God to and for His people.

This symbol is so universally recognized that it has also been used by many people, throughout history, to sign contracts or covenants. When sailors would join a crew on a ship, those who could not write would enter into their agreement by inscribing an X, their "mark." Even today, many forms that require a signature will denote the location to place your mark — or your signature — with an X.

Only later in history did the cross become a form of sacrifice that the Persians would invent and the Romans copied.

The Shape That Keeps on Giving

The tabernacle located at the center of the camp was constructed and organized in the same shape as the layout of the tribes. The cross formed by the placement and organization of the main accessories of the tabernacle was oriented in the same way as the camp, with the head of both crosses to the west and the feet to the east.

Diagram 6-3: The Cross of the Tabernacle

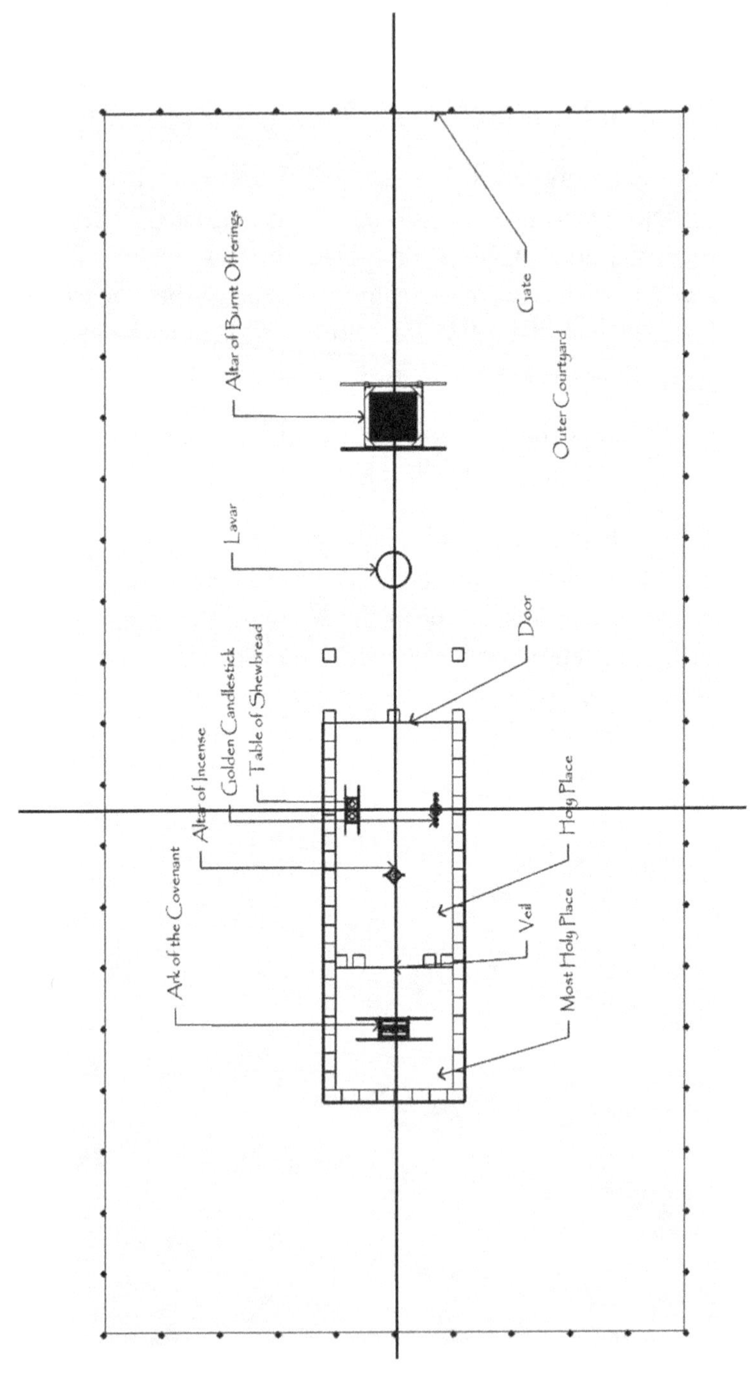

Each one of these four groups, of three tribes each, had a head tribe that was nearest the tabernacle. Their names and standards/ensigns were as follows:

Diagram 6-4: Tribal Names and Ensigns

Tribal Name	Standard/Ensign
Judah	Lion
Reuben	Man
Ephraim/Manasseh	Bull
Dan	Scorpion/Eagle

What Do Cherubim Have to Do with It?

It is interesting to note that Ezekiel 1 contains a description of the highest ranking angel in the kingdom of God.

> 4 As I looked, behold, a storm wind was coming from the north, a great cloud with fire flashing forth continually and a bright light around it, and in its midst something like glowing metal in the midst of the fire. 5 Within it there were figures resembling four living beings. And this was their appearance: they had human form. 6 Each of them had four faces and four wings. 7 Their legs were straight and their feet were like a calf's hoof, and they gleamed like burnished bronze. 8 Under their wings on their four sides were human hands. As for the faces and wings of the four of them, 9 their wings touched one another; their faces did not turn when they moved, each went straight forward. 10 As for the form of their faces, each had the face of a man; all four had the face of a lion on the right and the face of a bull on the left, and all four had the face of an eagle. 11 Such were their faces. Their wings were spread out above; each had two touching another being, and two covering their bodies. 12 And each went straight forward; wherever the spirit was about to go, they would go, without turning as they went.
>
> —Ezekiel 1:4–12

These cherubim are said to be around God's throne. They each have six wings and four faces. In verse 10 their heads are described as featuring the face of a man, a lion, a bull/ox, and an eagle. The cherubim are positioned around the throne, each one with the man's face in the front, with the lion's face to the right of that, the face of the bull/ox to the left, and the eagle's face behind. These cherubim faces are organized in the same fashion as the ensigns of the head tribes around the tabernacle.

The remaining verses of Ezekiel 1 reveal that these cherubim were surrounding the throne, which on earth would correspond to the Holy of Holies in the tabernacle. These same faces, or ensigns, were arranged around the tabernacle in the same sequence as the tribes of Israel. In addition, exactly the same creatures were oriented in the same way in the sky (more about that momentarily).

In other words, these same ensigns on the flagpoles of each of the four tribes were oriented in the same manner as the four faces on the four cherubim surrounding God's throne. In addition, these same ensigns/symbols appear overhead at night in the starry sky.

This cannot possibly be accidental. It could only occur by God's design.

Chapter Seven

Biblical Support and Use for Constellations/Zodiac

So — the twelve sons of Jacob, and the twelve tribes of Israel associated with each one, were represented by the constellations. And their camps were organized by God in a very specific way. But what are the stars themselves communicating?

Before we answer that question we need to address another one that usually comes up at this point. Some suggest that God would never use any pagan symbols. Stars, the sun, the moon, and other astrological signs sound way too demonic. Thus God would never use such things to communicate to mankind.

First of all we have to recognize that God finished creating His universe long before any pagan symbols of any kind whatsoever had been invented. However, Satan is a master of deception and perversion. History records his evil effects on truth. Everywhere we look it becomes obvious that the adversary has been very busy. He waters down truth and mixes it with all kinds of self-serving ideas that may help fool us into believing what he has to say. Whether it involves sexuality, morality, ideas about marriage, or dozens of other concepts, we continue to see Satan attempting to pervert God's truth in every way he can.

The same is true concerning the stars and constellations. Satan is not the Creator, whether in the physical world or in the arena of truth. What he is good at is mixing just enough lies with the truth so we will fall for the former and lose the blessing of the latter.

When it comes to what God created in the heavens, Satan has done his best to pervert the meaning and the stories that lie behind the starry host. Unfortunately, we sometimes spot the lie but then throw the baby out with the bathwater.

Perverted by Astrology

Astrology is one of the most familiar examples of this. Many modern believers correctly recognize it for what it is. It's a satanic deception that offers people so-called "advice about the future," based on information that supposedly comes from the stars. Yet because Satan has managed to fool so many others, those who recognize his lies often refuse to look a little deeper for actual truth.

Considering that God Himself created the heavens, is it possible that certain messages put there by God are now hidden from our view because of Satan's perversions? We believe that the answer to that question is a resounding yes.

In fact, many times in the Bible God has used heavenly bodies as signs to mankind. We have already noted that, in the first chapter of Genesis, God said that the purpose of the lights in the sky was for signs. In Matthew 2 God used the stars to guide the wise men to the newborn Messiah. In the book of Revelation, chapters 6, 8, and 9 (along with others), God uses signs in the heavens to forewarn us about the coming of events that will lead to the end of the age.

In Job 38:32, when God confronts Job, God refers to the *mazzarot* as something He created. But unfortunately, this Hebrew word is usually translated into English as "constellations," or left as "mazzarot." Its actual meaning is "the twelve signs of the zodiac."[12] The word *zodiac* comes from the Hebrew word *sodi*, which means "the way."

Many think that "the way" being referred to here is the way of the sun. It is true that, during the course of the year, from the perspective of a viewer on the earth, the sun travels through each one of the constellations. Yet I believe that calling this "the way," as referred to in Job 38:32, is yet another deception put in front of us by the adversary. Yes, the sun follows its own "way." But that's not what the book of Job is talking about! The constellations also tell stories and deliver prophecies. Or as Genesis 1 informs us, they reveal to us *appointed times* and *signs foretelling of future events.*

On the other hand, in the Bible's Hebraic way of doing things, God could actually be implying something about Himself. The word in Hebrew for "the sun" is *shamash,* which many times refers to God Himself because God is the one who provides the light that allows us to find our way in life. So, what God could be saying here, by calling the "sodi" or "the way of the sun through the constellations," is that he is revealing to us HIS way — the part of the plan that HE is promising to fulfill.

These future events speak to us about the very plans of God that have unfolded before mankind's eyes over the centuries. They also speak of those that will come to their culmination in the years immediately ahead. No wonder Satan wants to hide this truth from us. The last thing he wants is for men and women to be aware that God has a master plan that is being fulfilled. This would only confirm the very truth that Satan desperately wants to hide from us.

In spite of all that, as we ourselves become aware of the "voices" — or plans — revealed by the stars in the sky, it will become very clear what "The Way" actually refers to. The Bible talks about the way that we should walk. Yeshua even tells us that He is The Way, He is the standard for the way we conduct our lives. The *sodi*, or the way of the sun, is actually the revelation of the plan of God to redeem mankind from their sin.

The constellations tell of this way, with the "flow of the narrative" deriving its cohesion by following the order in which the sun travels through the constellations.

But more on that later. The most important thing I want to convey here is that I am in no way advocating the study, belief, or practice of astrology. It is pagan in all of its machinations and manifestations, and is sourced in the minds of devils.

Instead, what we want to do is to try to discover what God originally gave to mankind, via the stars, as an honest guide for our lives. I believe we will discover that the constellations were given to us to encourage us and give us hope in times of trouble and desperation. In the days ahead mankind will need as much truth as he can get to help keep his attention steady and true, to stay the course.

The messages in the sky are not about our love lives and our personal strengths and weaknesses. They are about the love that God has for us. It is so great that He sent a Redeemer to rescue us from ourselves. The message of the stars is about our God, our Messiah, and our Groom.

And therefore it's about us.

Chapter Eight

Ancient Belief in the Constellations

The constellations are recognized all over the world. But this is not a modern phenomenon. Most ancient cultures were aware of their existence, so they studied them and sometimes even worshiped them. Cultures as diverse as those of the Chinese, the Babylonians, the Egyptians, the Greeks, the Romans, and the Sumerians, to name just a few, all had firsthand knowledge concerning the constellations.

And yet, all these ancient, widely dispersed societies recognized the same twelve constellations, and all their names for them meant the same thing. In fact, each culture has pictured each of these twelve constellations in ways too similar to have come about by accident. Such pictures include universal images of snakes, kings, and great men and women.

Some observers suggest that the reason for the similarity between representations of the constellations is that the people of each culture saw the same outlines formed by each set of stars. The problem with that proposal is that the stars in each constellation do not form or outline the images. It is as though an artist took a clear plastic sheet, drew twelve figures, oriented them in a large circle, then placed the clear sheet on top of a picture of the *ecliptic*, which is the path that the sun follows through the sky over the course of the year, relative to the stars, as seen from the vantage point of the earth.[13]

But the sheet with the images painted on it would not in any way be painted or organized to reflect any particular placement of the stars. It is as though the two pictures, one of the stars and one of the constellations, were randomly placed on top of each other. In other words, when one looks up at the stars in the night sky, their organization does not suggest any image — and certainly not the images that have been imposed on them from ancient times.

A Common Source

This uniformity of the constellation "pictures," being agreed on in this same fashion in ancient cultures all over the world, is truly remarkable. At a minimum this suggests that these ancient versions of what the constellations represent pictorially, which have persisted throughout the centuries, all had a common source.

So, if they did have a common source, what does that suggest about the constellations' worshipers or admirers? Did all these people migrate from a single locale where they were exposed — jointly —to a single set of understandings?

Is it just accidental that this migration, from one common location, is exactly what the Bible suggests? Are so-called fables about Adam and Eve and the Tower of Babel supported by evidence, after all? All of this certainly presents a powerful argument for the idea that God created the constellations — and the rest of the universe as well — exactly as the Bible teaches.

It also implies that the ancient ideas associated with the constellations are not manmade. How could uniformly recognized constellations that did not correlate visually with their starry backgrounds be invented in all these different societies? And at the same time be given similar names?

Three Decans Each

Most of us are aware of the twelve major constellations, but few of us realize that there are at least thirty-six others. The twelve major constellations each have three other minor constellations associated with them. These thirty-six additional constellations are called *decans,* which is pronounced exactly the same as *deacons.*

In a different context we know that deacons, in this spelling, are people who serve others; they complement what others are doing. We see this in worship services when the leader conducting the service is assisted by deacons who handle many of the details so the service can be run smoothly.

These starry decans do the same thing. Since the constellations are telling a story, the decans embellish the message by providing background and completion to the theme of the lead constellation of which they are a part.

Is it a Decan or a Deacon?

Our word *decan* has very ancient roots. The Hebrew word *deka,* spelled *dalet, kaf, aleph,* seems to be the source word. This word means "to break into pieces or into a small piece." This would help to verify that decan is one of many English words that find their origin in the ancient Hebrew language.

Indeed, many modern scholars are coming to the conclusion that Hebrew was the original language from which all other languages evolved. This, of course, is no surprise to ancient Jewish scribes who were students of the Tanakh. From the earliest times they proposed that the language with which God spoke the universe into existence was Hebrew.

However, at the Tower of Babel God intentionally created all the other languages from which every language spoken today has evolved. When linguists track our current languages back in time, in geographical terms they end up in Iraq, the exact location of the Tower of Babel. Most Hebrew scholars also propose that God will restore the knowledge of this original language when He comes back. God refers to this language as "pure" in Zephaniah 3:9.

The migrations of people groups over the globe all began in the Middle East. Even the origins of domesticated grapevines all lead us back to Turkey. Remember, Noah planted the first grape vineyard after the Flood.

Circular Reasoning, Anyone?

These twelve primary constellations each have three decans, for a total of forty-eight constellations. All of these pictures have very ancient but ubiquitous origins. They are oriented in the sky in a large circle, around the north star, which revolves overhead while we lay in bed each night.

Like all other circles, the circle we just mentioned is commonly divided into twelve 30-degree sections, totaling (again, like all other circles) 360 degrees. This practice also has very ancient origins. It seems that the ancient Sumerians, Akkadians, and Babylonians, living in the years preceding 2000 BC, all had lunar calendars. This meant that their years, like ours, were divided into twelve 30-day months, and the sun's annual path across the sky took 360 days. Or, to be more correct, the earth's passage through all twelve "sections" of its circular path around the sun took 360 days.

Late in the 8th century BC, most of the civilizations existing at that time modified their calendars by adding about five days to the earth's annual circuit around the sun. Some researchers today believe that an astronomical event occurred — perhaps some interaction between Mars and the earth — which had the effect of slowing down the speed of the earth and increasing the speed of Mars, thus causing the earth's travel around the sun to take an additional five days or so.

Where Does Astrology Fit into the REAL Picture?

In ancient civilizations, each section of 30 degrees of the earth's orbit around the sun represented the distance the moon appeared to move through the stars and the twelve constellations each month, from our earthly perspective. Modern astrologers still believe that these 30-degree sections represent the amount of time that the moon still spends in each section, which the early astrologers called "houses."

Unfortunately, these houses, which are known to astrologers by the names of the twelve constellations (and are also called the "signs of the zodiac"), are all assigned thirty degrees of orbital travel even though the amount of such travel involved in each one actually varies greatly. Therefore they occupy different-sized portions of the complete circle, which highlights a huge logical fallacy with respect to the rigid 30-day time frame each "sign of the zodiac" is given.

How can the claimed relationships between the zodiacal "signs" of our birth be considered significant when millions of us have not been born during what might appear (from our perspective) to be the moon's travel through a particular zodiac, but actually could be keyed to that so-called travel through one of the zodiacs traversed either before or after the one during which we were supposedly born?

Virgo Was the First

Because these constellations, from ancient times, have been depicted — both pictorially and conceptually — as being arranged in a circle, there has been some confusion as to which one should be the starting point. The key to this question has been discovered in Egypt. The temple at Dendera, in a carving in the temple walls, depicted the sphinx between Leo and Virgo. The woman's head of the Sphinx represented Virgo, the virgin.

The constellation Virgo was the first picture and the beginning of the stories in the sky. The back of the Sphinx was a lion and represented the last of the constellations in the circle: Leo. However, astrologers begin with Aries because their assumptions are steeped in paganism rather than history, and they ignore archeological finds. This juxtaposition of the sphinx strongly suggests that the starting point of the circle of constellations should actually be Virgo.

Jewish scholars, aware of the existence of the twelve constellations, believed that each one of these twelve correlated to the twelve tribes (sons) of Israel. Thus, on each of the standards of the tribes that traveled across the wilderness on their way to the promised land, they positioned twelve symbols, each of which represented one of the twelve constellations.

This concept was further embellished by the Hebrew understanding of the number twelve. To them it meant "completed righteous government." Thus, as we develop our understanding of the stories the constellations speak of we begin to recognize that they tell of a future time when this world will be completely restored and His kingdom will reign.

And surely it will also be righteous.

Chapter Nine

The Construction of the Zodiac

Before we launch into a description of the twelve constellations, each with their three decans, let's establish a clear understanding of how they appear in the sky. To do that we need to understand how they are associated with the sun as the earth rotates and the sun appears to move through the sky during the day. And, we need to consider how the sun appears to move through the constellations through the course of the year.

Let's begin with a simple example rather than trying to describe the real thing. By putting ourselves in the middle of a round room we can better understand these astronomical associations. In such an example, the spot we are in is the same as the location of the sun in our solar system. The earth revolves around the sun once a year in a counterclockwise direction (at least from the perspective above the north pole). The earth also rotates on its axis in a counterclockwise fashion. To further complicate things, we know that the axis on which the earth spins is not perpendicular to the plane of the orbit of the earth.

In other words, if a line existed that connected the sun to our earth, our axis, if it were perpendicular to the plane of its orbit, would be at a 90° angle to this imaginary line during the earth's entire circuit

Diagram 9-1: The Tilt of the Earth[14]

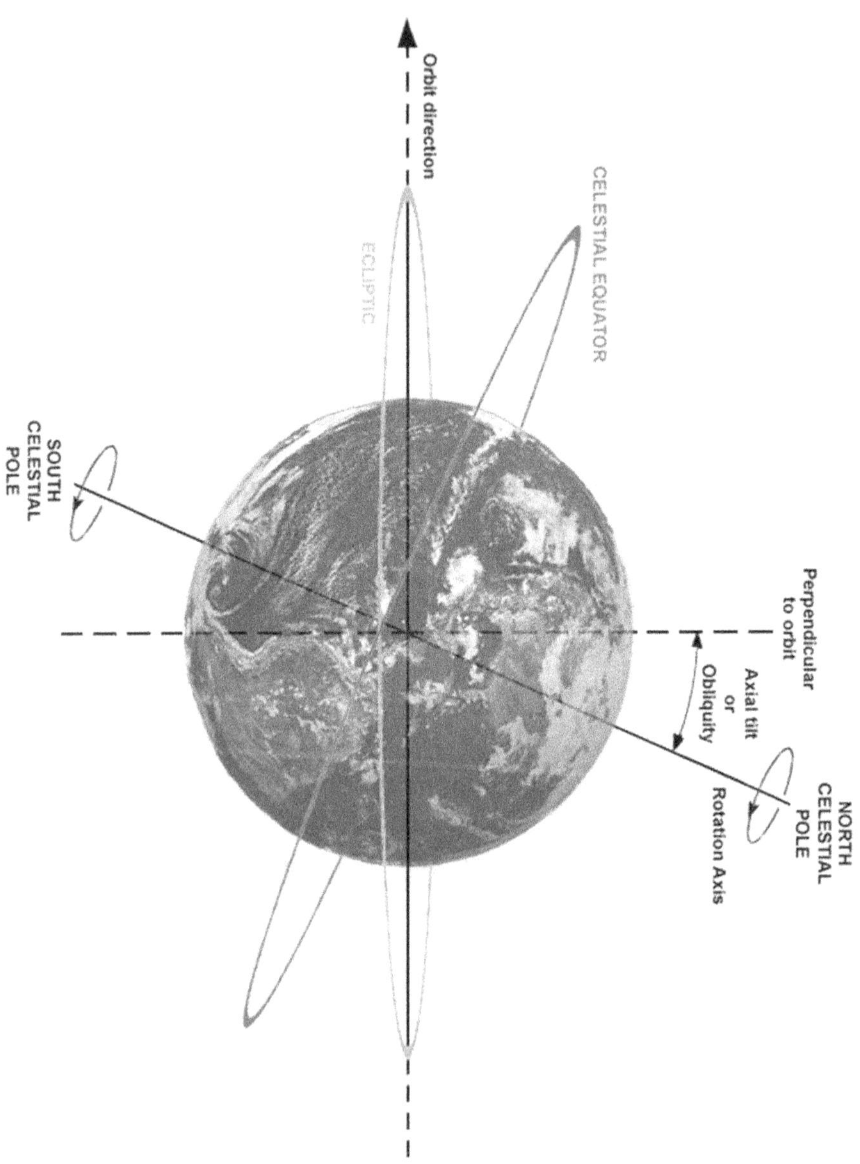

around the sun. But the earth's tilt on its axis is *not* perpendicular to its orbit around the sun. Instead it tilts 23.44° *away* from 90° (see Diagram 9-1).

Given all that, if you were standing on the sun, when viewing the earth in the spring you would see that the axis of the spin of the earth was tilted a bit toward the right, away from the earth's direction of travel. This would be the situation as it would exist on March 21, at the spring equinox.

However, as the earth continues revolving around the sun in a counterclockwise fashion, on June 21 the axis would now be leaning directly toward you and the sun, instead of appearing to lean a bit to the right. On this date we celebrate the summer solstice, and in the northern hemisphere we enjoy the longest day of the year.

Then, in the fall, on September 21 the earth would have made another complete quarter of its orbit around the sun and would now be exactly opposite its spring location. And, of course, the earth's tilt on its axis would still be the same as at the spring equinox, except that now the earth would be leaning to the left of you. This day is known as the fall equinox.

Finally, at the start of winter on December 21, known as the winter solstice, those living in the northern hemisphere experience the shortest day — and the longest night — of the entire year. From the viewpoint of the sun the earth's tilt would now be pointing away from the sun. Then, at the spring equinox (March 21) the earth would complete its orbit around the sun and would arrive back at the same location it was in when we began this illustration.

Ready for the Stars

Now that we understand the sun and the earth's orientation to each other, let's add in the stars. In our solar system example, pretend that we are still standing on the sun, in the middle of a circular room, with the earth orbiting around us in a counterclockwise fashion. If we were to keep our eyes on the earth we would also be able to see the wall behind it as the earth moved during the year. So, on this wall let's now place all the stars that exist in the sky above as they would properly appear during the year. Then, if we kept the earth in the center of our line of sight, as the earth orbited around the sun we would see different stars appearing behind the earth as it traveled.

Now let's change our location from being near the sun to standing on the earth. The first thing we will notice is that it is a lot cooler. Beyond that, as the earth turns on its axis, as we look up into the sky we would find ourselves facing different walls throughout the day, while a whole new set of walls would come into view during the night.

Nighttime, of course, would occur when we were not facing the sun. That's when, via time-lapse photography, we can appear to see the stars streaking across the sky. But actually the stars are not moving at all. They appear to move across the sky because the earth is spinning, even as the sun appears to move across the sky during the daytime for the same reason. Conversely, if we could see the stars during the day, the stars that we would see would actually be on the wall of our circular room on the other side of the sun.

Star Talk

Now, what's the point of all this star talk? Well, at each time of the year, whether it might be spring, summer, fall, or winter, different stars would appear at night. In fact, the very stars we cannot see during the day in the spring are the very ones we see at night in the fall. Likewise, the stars we cannot see during the day in the summer are the same ones we would see at night in the winter.

Now, let's add in one more important set of facts. As we have already explained, the stars themselves are grouped into twelve constellations, each with their three decans. Going back to our "round room" example above, as the earth revolved around the sun the stars would show up on our walls in exactly the same order each year, with the first constellation, Virgo, starting the annual cycle and the last constellation, Leo, completing it. Because of this, Virgo and Leo are next to each other. So, as the progression ends with Leo overhead at night, and the new year begins, Virgo "replaces" Leo in that dominant position at midnight.

Thus *Virgo* is directly overhead at midnight during the spring equinox on March 21, just as *Pisces* is directly overhead at midnight during the fall equinox in September. However, as the night progressed beyond midnight and the earth continued to rotate on its axis, by the time it had completed a quarter-turn, the first rays of the sun would begin to appear.

But beginning in the early hours of the night we would have seen a progression of constellations appearing as follows: Virgo overhead at midnight, then Libra, Scorpio, Sagittarius, and Capricorn. Then, just before the sun came back into view we would see Aquarius rising in the eastern skyline.

Of course, if the sun did not wash out the light from the stars during the day we would see Pisces appearing to rise with the sun, followed by Aries, Taurus, Gemini, Cancer, and then Leo, all rising from the eastern skyline during the day. Finally, again at midnight in March, Virgo would take its usual position after rising from the eastern skyline about three hours earlier.

One Overhead and One Rising

One more thing might not be obvious at this point. In the fall, all of the above is just the opposite. Pisces would be overhead at midnight, followed by Aries, Taurus, Gemini, Cancer, and then Leo. Then, at dawn, Virgo would rise with the sun and Libra, Scorpio, Sagittarius, Capricorn, and Aquarius would follow.

In other words, during each month we see a different constellation overhead at midnight, and a different one rising with the sun at dawn. The order of these overhead-at-midnight appearances is the same order of their appearances noted above. Consequently, if we could see the stars during the daytime, during each twenty-four-hour period we would see a complete parade of twelve constellations, each with their three decans, passing overhead.

And each month a different constellation would appear directly above us at midnight, with a different one rising with the sun. This "new" constellation, would be the next one in the order above.

Chapter Ten

The Story in the Sky: Act One

In Genesis 1, when describing the work God did on day four, He informed us that He made the lights in the sky to foretell and forewarn us about things to come. Psalm 19 then expands on Genesis by telling us that the heavens have voices. And, that they pour forth knowledge. In verse four God claims that the luminaries in the sky can be heard wherever mankind might exist:

> Their line has gone out through all the earth,
> And their utterances to the end of the world.
>
> —Psalm 19:4

Now, if we believe the claims of God then this passage should give us pause, because in our day the only thing we hear that "comes from the heavens" are the dubious predictions from astrologers. Their prophecies are built on the notion that every person's daily life is influenced by the movements and orientation of the stars, planets, moon, and sun on that particular day.

Certainly these voices cannot be what God was talking about. We all know that these "foretellers" usually claim to get their messages from powers and beings that do not include the likes of Yahweh. Their messages have nothing to do with declaring the glory and handiwork of God (Psalm 19:1). They glorify the creation rather than its Creator.

So, if the foretellings of these soothsayers are not the voices heard coming from the stars that God put there, what is? Let's see if we can wade through all of the myths that are now associated with these constellations, by stripping away the false voices and finding the true words that were originally given to us by God. We have learned that He put the stars in the sky to warn us about things He wants us to be aware of. He wanted us to watch for the fulfillment of His messages. Exactly what are those "messages" all about?

No One Else Is 100% Consistent

Students of the Bible know that God's Word is full of prophecies about the end of time. The study of this subject has even been given a name to differentiate it from other biblical topics: eschatology.

Knowing that our God is one hundred percent consistent with truth (truth does not change over time, nor does it change from culture to culture), and also knowing that what is important to Him should also be important to us, we should expect the voices in the sky to be similar to the words He has given us in the Bible. In fact, the messages in both should complement each other if they are coming from the same person — namely, Yahweh.

For example, the Bible does tell about the creation and the fall of mankind as a result of Adam and Eve's sin. But it also tells about a great hope for mankind — a hope that includes the coming of a Lamb who will pay the price for our sins by dying on a cross in the hope that mankind will return to his first love.

But that is not the end of the story, for the Bible also explains that someday that Lamb will come again, not as a Lamb this time but as a King. He will deal with the dragon, the one who caused Adam and Eve to fall in the first place, and who continues to deceive us so that we'll do the same. This coming confrontation will result in God's reclaiming the earth, setting up His kingdom, and restoring believing mankind to a right relationship with Him.

This is the hope that we find in God's Word, referred to as the "Good News" in the text of the Bible. This good news includes new restored bodies and lives for those who will rule and reign forever with our Creator here on earth. Thankfully, the power and influence of the dragon will be removed from the earth and from man for all time. This snake and his deceptions will be thrown into the Lake of Fire.

Good News for All!

We are often struck with the New-Testament topic concerning "Good News." In our churches this "Good News" is sometimes presented in a way that implies that it is something new; maybe even something that was not expected. What is this good news? And is it something never spoken about before?

First of all, this good news is all about the coming of the Messiah, Yeshua. But more important, it tells us that His accomplishments during His life here on earth have brought us salvation through faith. As Paul explained:

> . . . if you confess with your mouth Jesus as Lord, and believe in your heart that God raised Him from the dead, you will be saved; [10] for with the heart a person believes, resulting in righteousness, and with the mouth he confesses, resulting in salvation.
>
> —Romans 10:9–10

What is missing from many pastoral sermons, however, is that the "good news" spoken about in Romans and many other places in the New Testament is actually a continuation of a similar theme in the Old Testament. In fact, the Romans 10 passage that refers to this good news, in verse 15, is quoting Isaiah 52:7 as revealed below:

> How will they preach unless they are sent? Just as it is written, "How beautiful are the feet of those who bring good news of good things!"
>
> —Romans 10:15

> How lovely on the mountains are the feet of him who brings good news, Who announces peace and brings good news of happiness, Who announces salvation, and says to Zion, "Your God reigns!"
>
> —Isaiah 52:7

Here in Isaiah the original Hebrew word for "good news" is *basar*, which usually gets translated as "tidings." This is the Hebrew source word for the Greek word we find in the Septuagint and the New Testament for "good news."

Unfortunately, the translators are not consistent with their English word choices and blur this continuity. From ancient times God has given mankind hope: That He would come someday and take away the sins of the world. This has been the good news for all mankind, from Genesis to the book of Revelation.

Written in the Sky

Before the Bible existed, did God offer this same hope to those living prior to 1500 BC, which is when Moses, via God, wrote his portions of the biblical text? I think so. Every night the ancients could look up and both recognize and comprehend the messages in the heavens. We're going to find out that the very hope that modern man relies on for his salvation and restoration was given first to ancient man. It was written in the sky.

Several excellent books go into great detail about these constellations. I have included a list of them in the bibliography at the end of this book. The stars and their names, as well as various particulars about each one, are described. Certainly, these works should be read by anyone desiring more information.

However, I won't repeat all of their insights but will include only a general description, followed by my own thoughts. My focus here is to discover the plan of God as it was revealed to mankind through these constellations, long before the Bible was written. I'd like to compare it to the goals, purposes, and designs of God, as revealed in His Word, and see how they match up. Then, by looking at all this through the Hebrew lens, I hope to uncover additional insights that the voices from the heavens are proclaiming.

As we approach these twelve constellations, each with their three decans, one additional point needs to be made. As explained previously, it is accepted by many that these twelve are organized into three groups of four constellations each. Because these pictures in the sky tell a story, we will call the first set of four constellations, each with their three decans, Act One. The next chapter in this book will contain the second four, in Act Two, followed by the last group of four constellations, which will be Act Three.

The Constellation Virgo and its Decans

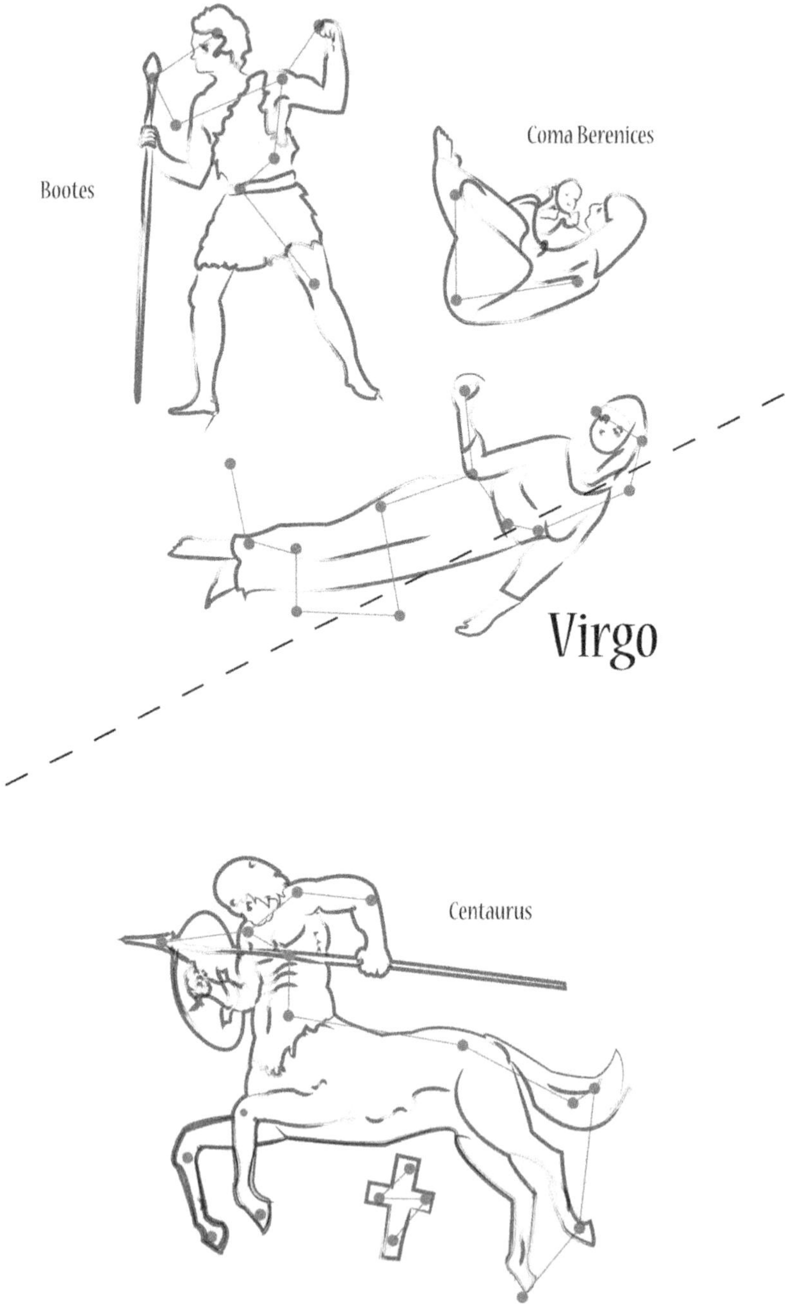

Coma Berenices

Bootes

Virgo

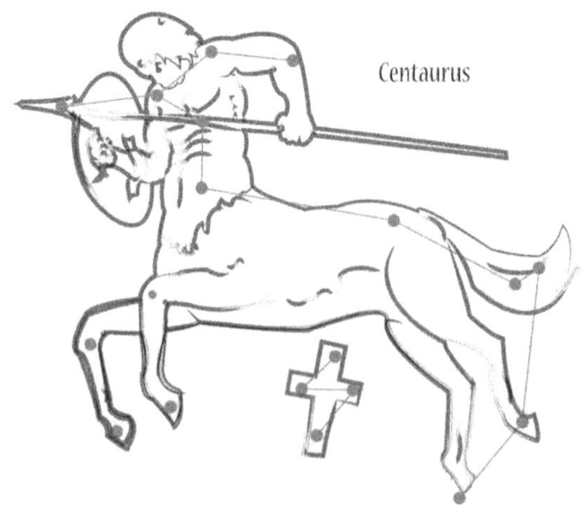

Act One, Constellation Group One: Virgo

If our speculation about God's story being expressed in the starry host is correct, Virgo should begin the conversation about His plan. If so, this first constellation would have given hope to the ancients, which is exactly what Virgo does. It and its three decans provide a general overview of the plan of God, and give those who discern that overview plenty of reason for hope.

Remember, the ancients were very aware that they were suffering under the curse of their sins. God had cursed the ground, requiring man to struggle and battle the ground to force it to bring forth food to meet each family's daily needs. Women knew very well what that terrible decision at the tree of life meant. All the pain that now came at childbirth reminded them of the consequences of the curse of death.

What a contrast life in the Garden of Eden was compared with this new life! I'm sure the question in their minds must have been, "Is there any way we can get back to the way it was?" This first constellation answered that question. It gave them hope, but not hope in themselves. Their restoration lay in the work of another; nothing they could do on their own would mend the damage.

A Virgin Shall Conceive

Virgo's first decan is *Coma*. Its name means the "Desired One." It features a woman, sitting in a chair and holding a child in her arms. This constellation told the ancients that someday a virgin would give birth to a son. He would be very much desired and would hold much promise.

Centauras, fashioned in the form of a centaur (half-horse, with the head of a man), follows Coma. This decan tells more about the virgin's son. It proclaims that, even though he will be the desired one, he will also be despised. As Isaiah 53:3 says:

> He was despised and forsaken of men, a man of sorrows and acquainted with grief; and like one from whom men hide their face He was despised, and we did not esteem Him.

In Hebrew, the word translated as "despised" is *bazah* and is the Hebrew name for this decan. Centauras is located just above another decan, called the cross, which is a member of another group of constellations. The suggestion here is that this son will be despised and rejected and will suffer a terrible death by crucifixion.

The back portions of Centauras are in the form of a horse, while its front, as indicated above, takes the form of a man. This two-part aspect suggests that this coming son will have two natures, one of man and one of something else. He is holding a long spear, readying it to impale another decan called the *Victim*. This suggests that when this son comes he will be engaged in constant conflict. There will be a fierce battle, and we will learn more about the son's opponent as we go along.

What a terrible story so far. A virgin will come, bearing a son. This is a miracle in itself. That event should certainly catch man's attention and mark the child as someone special. However, he will be desired by mankind but at no benefit to himself, for his life will be taken from him though his death on the cross.

Behold! He Comes!

But wait! One more decan remains in this cluster of four. It is *Bootés*, which means "He comes!" The Bible speaks many times about the coming of God back to the earth. Biblical chapters such as Acts 1 and Revelation 1 and 19 come to mind. Revelation 1:7 brings this point home:

> BEHOLD, HE IS COMING WITH THE CLOUDS, and every eye will see Him, even those who pierced Him; and all the tribes of the earth will mourn over Him. So it is to be. Amen.

This Bootés is a picture of a man standing and holding a rod in his right hand and a sickle in his left. Revelation 12:5 also speaks about a son who will come back to earth and will rule with a rod of iron:

> And she gave birth to a son, a male child, who is to rule all the nations with a rod of iron; and her child was caught up to God and to His throne.

Certainly, Revelation 14:14–16 also correlates with the message given by this decan:

> [14] Then I looked, and behold, a white cloud, and sitting on the cloud was one like a son of man, having a golden crown on His head and a sharp sickle in His hand. [15] And another angel came out of the temple, crying out with a loud voice to Him who sat on the cloud, "Put in your sickle and reap, for the hour to reap has come, because the harvest of the earth is ripe." [16] Then He who sat on the cloud swung His sickle over the earth, and the earth was reaped.

What the stars seem to be revealing is that this son, the one who was despised, rejected, and even killed, will return. But this time He will not come to die but to rule. He will harvest the earth and set up His kingdom.

The story is now complete. And yes, it is brief with not a lot of detail. But we have the basics revealed to us by this first set of constellations, the head of which is Virgo.

- A virgin will come bearing a son.
- There will be something different about this man child, for he will be a man but he will also be something else: something kingly, something royal, something strong.
- He will be sought after but will also be despised.
- He will suffer death on the cross, at which point the contrast will be obvious for those who watch. An innocent man, with great authority and strength, will suffer a criminal's death.
- He will then come back again, but this time as a conqueror wielding great authority.

This revelation is very similar to the one detailed in Genesis 3. As you will recall, God cursed the snake after Satan deceived Adam and Eve. God then revealed that a conflict would come. The battle would be between a future offspring of Eve (Yeshua) and Satan himself. And, in the end, Eve's offspring would win by giving Satan a deadly head wound, but Yeshua would suffer death on the cross in the process.

In Genesis 3 God kills a lamb to make a covering for the first two people He created, who are now fallen from their perfect state. Like the story in the sky described by the constellation Virgo and its decans, this lamb metaphorically presents an interesting idea. There is a price to be paid for sin, but only someone who is perfect can step forward and pay the price on the trespassers' behalf. Unfortunately, the price tag for sin is extremely high, for it requires death.

What Is It About Blood?

We often think that the shed blood saves us from our sins. However, it is not actually the blood itself that makes the difference — it's the life or soul *within* the blood. Leviticus 17:11 states:

> "For the life of the flesh is in the blood, and I have given it to you on the altar to make atonement for your souls; for it is the blood by reason of the life that makes atonement."

Thus Yeshua gave up His life's blood on the cross on our behalf. The Hebrew word in the above verse for *life* and for *soul* is the same. It is the word *nephesh*. This word is most frequently translated as "soul," but can also be rendered "life" and "creature." Its first occurrence comes in Genesis 1:21:

> God created the great sea monsters and every living *creature* that moves, with which the waters swarmed after their kind, and every winged bird after its kind; and God saw that it was good. [Italics added.]

Nephesh is the underlying Hebrew word translated above as "creature." That is certainly a correct choice to describe various living animals. However, let's take a look at another verse in Genesis 2:7:

> Then the LORD God formed man of dust from the ground, and breathed into his nostrils the breath of life; and man became a living being.

This verse describes the creation of man rather than animals. The underlying Hebrew word for "being" is the same word, *nephesh*, used back in Genesis 1:21 where it was translated as "creature."

Unfortunately, readers do not get a clear understanding of what God is trying to tell us because the translators are being inconsistent. What God is conveying in both of these verses is that He put a soul in all living creatures, including mankind. It is this very soul that must be given up as payment for sin.

The difference between man and the animals does not reside in the type of souls but in how God created them. Animal souls were created by the spoken word of God. They became living by the very words that came out of His mouth. On the other hand, man became a living soul through the breath of life that came from the mouth of God.

The Hebrew words for "breath of life" are *neshmah chai*. *Neshmah* is a word that means "Spirit of God" in Hebrew. So this verse supports the rabbinical idea that a bit of God Himself dwells within each person.

Thus the difference is relatively simple. God created the animal souls by His spoken word, but He created mankind's souls by placing a bit of Himself in each person. God was making a very big point here, which is commonly missed. Man is very special to God. Yes, he is much more important than the animals, although God loves and cares for them as well. That's why it is our job to manage and protect them.

Should Doctors Read the Bible?

Several centuries ago it was a common medical practice to "bleed" people who were suffering from various types of ailments. As with so many other false scientific concepts that have existed over the last thousand years, if the same people had read the two biblical texts in the previous text box they would have realized that the very last thing you want to do to a person — especially a sick one — is "bleed them out." Today we know of many reasons why this was a very deleterious medical practice. But even without today's knowledge, God embedded in His Word the knowledge and instruction that could have saved a lot of lives. If only we would be students of the Bible.

The Constellation Libra and its Decans

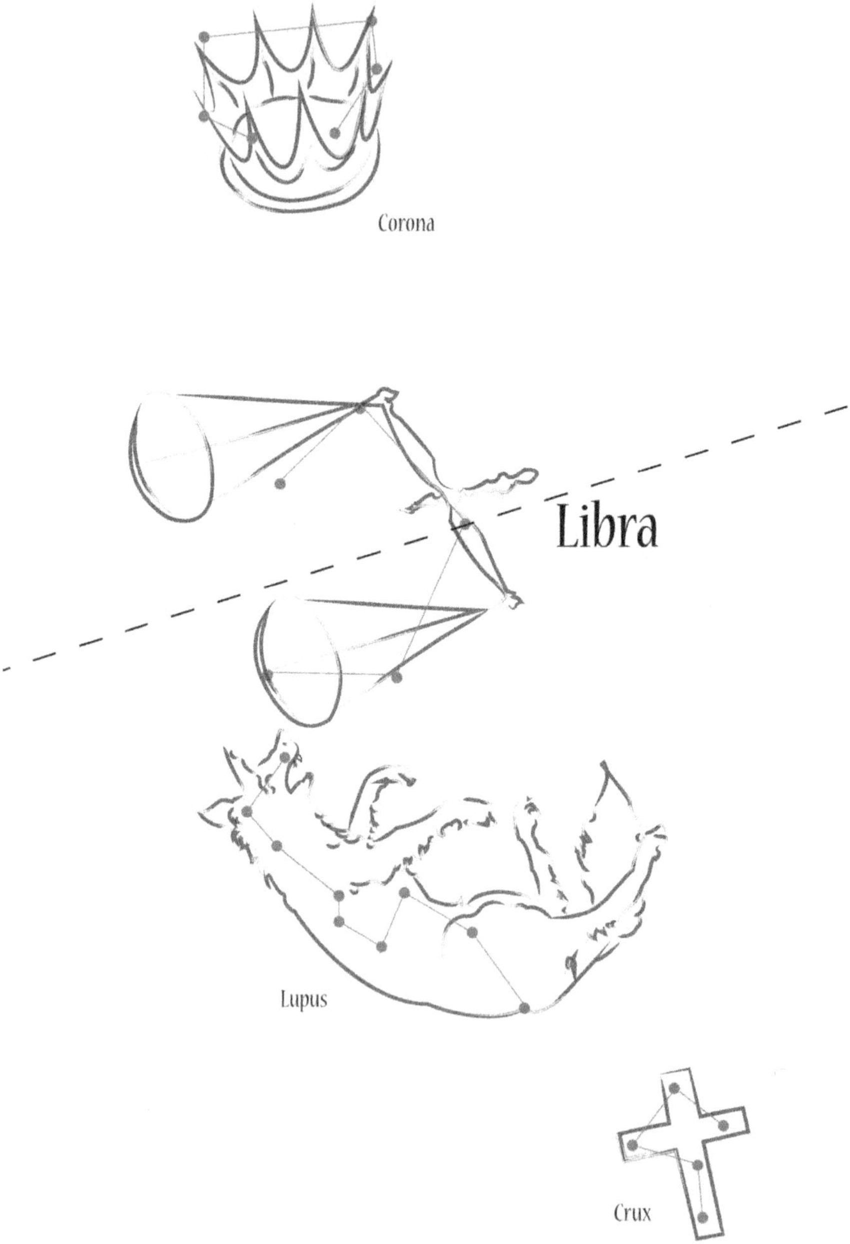

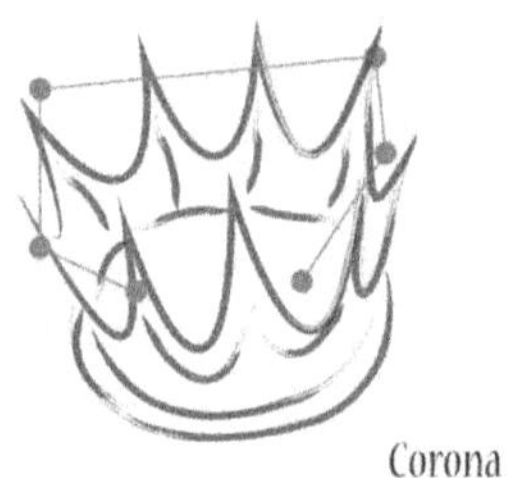

Libra

Lupus

Crux

Act One, Constellation Group Two: Libra

Libra is overhead at midnight during the month of May. It follows Virgo's similar appearance in late March or April. Libra, along with its decans *Crux*, *Lupus*, and *Corona*, rises with the sun in November, whereas Virgo rises with the sun in October.

This constellation is depicted by the picture of a scale. It's not one of those newfangled digital scales, but an old-fashioned device as sometimes depicted on court buildings. There the scale represents the measurement of law and justice, which should be equal — at least theoretically.

Our lives sometimes come face-to-face with the law. At such times our choices are measured against the law and what would be the proper penalty for any violation. The starry scale reveals that mankind's weight is found wanting. (And no, don't take this as permission to ask for another helping of dessert.)

That is exactly what God told Belshazzar when the king of Babylon drank from the cups from the temple of God. A hand appeared in the banquet room where the king had called for the temple implements to be used in the celebrations of that fateful night. A disembodied hand appeared and wrote the words 'MENE, MENE, TEKEL, UPHARSIN' in front of the guests, astonishing and frightening them all.

The prophet Daniel interpreted the writing starting in Daniel chapter 5:26–28:

> "This is the interpretation of the message: 'MENE' — God has numbered your kingdom and put an end to it. 'TEKEL' — you have been weighed on the scales and found deficient. 'PERES' — your kingdom has been divided and given over to the Medes and Persians."

The Hebrews called Libra *Mozanaim*. The Hebrew word above, for "scales," is also *mozan*,[15] which is the same word in its singular form. Unfortunately for mankind, we do not measure up either, collectively or individually. When our thoughts and actions are put on the scale of justice we all fall short of the standard. And this standard, by the way, is perfection — mankind's original state of being.

We would also suffer the same fate as Belshazzar if not for the Lamb's payment, which was effectively placed on the scale for all mankind. It is the only weight that brings balance back into the universe and into our lives, providing we accept and embrace it.

> [11] Then I saw a great white throne and Him who sat upon it, from whose presence earth and heaven fled away, and no place was found for them. [12] And I saw the dead, the great and the small, standing before the throne, and books were opened; and another book was opened, which is the book of life; and the dead were judged from the things which were written in the books, according to their deeds. [13] And the sea gave up the dead which were in it, and death and Hades gave up the dead which were in them; and they were judged, every one of them according to their deeds. [14] Then death and Hades were thrown into the lake of fire. This is the second death, the lake of fire. [15] And if anyone's name was not found written in the book of life, he was thrown into the lake of fire.
>
> —Revelation 19:11–15

The Price of the Shortfall

The next two decans further elaborate on our shortfall by describing the price that was paid, who would pay it, and the means by which it would be paid. Of these two decans, Crux is the first decan within the constellation Libra. Crux is the Latin word for "cross." We learned about this decan earlier when it was found to be just under the decan Centaurus in the constellation Virgo. As we explained earlier, Centaurus is the two-natured centaur; half man, half horse. Centaurus represented the two natures of God. When He came as a child He presented Himself first as a member of the human race, but also as the God who created the universe (John 1:1-9).

The decan Crux is positioned directly underneath this centaur, suggesting the price He will pay on our behalf.

The second decan, *Victima* (which is also known as Lupus), follows Crux. Its Latin meaning is "victim," and in some ancient drawings it appears to be a lamb. It is a picture of a four-legged animal that has been killed by the point of the spear held by Centaurus. Both Victima and Centaurus, the victim and the centaur, represent God. So the story conveys the idea that what actually killed the son who came from Virgo, the virgin, was God Himself.

Now — how could God be both the victim and the perpetrator? The common accusation within the church community today is that the Jews were responsible for the death of our Lord. However, a closer look at the biblical text, augmented by understanding the legal realities of the day, reveals that the Jews did not possess the right to punish people in such a way. Only the Romans, who were Gentiles, had the authority to take away a human life. But these two decans remind us of another verse, often overlooked or ignored in this particular context, which says:

> For God so loved the world, that he gave his only begotten Son, that whoever believes in Him shall not perish, but have eternal life.
>
> —John 3:16

God gave Himself willingly to pay for mankind's lusts. Neither the Jews' nor the Gentiles' decision resulted in the death of our Lord. And more important, without that one death, we ourselves would be dead. So when we look for someone to blame for God's suffering, let's remember that it was His will to come and die. And without His payment we would have to pay it ourselves. The decans are conveying the willingness of God to offer up Himself as the Victim and the payment for our sins. And He was also the centaur: by His hand our God and King died.

It Doesn't End There

Crux and Victima suggest that someone, represented as a lamb, will come and offer up Himself on a cross to die. But the story of this constellation does not end there. The third decan is Corona Borealis, better known just as Corona. The full name is Latin for the "northern crown." Corona is a small decan in the northern sky, and its association with the constellation Libra suggests that the reward given to Victima, for the price He was willing to pay, was this crown.

Of course we are all aware that this Victim is portrayed as Yeshua in the Bible, and He will appear again wearing this same crown.

> [5] And one of the elders said to me, "Stop weeping; behold, the Lion that is from the tribe of Judah, the Root of David, has overcome so as to open the book and its seven seals." [6] And I saw between the throne (with the four living creatures) and the elders a Lamb standing, as if slain, having seven horns and seven eyes, which are the seven Spirits of God, sent out into all the earth.
>
> —Revelation 5:5, 6

> Then I looked, and behold, a white cloud, and sitting on the cloud was one like a son of man, having a golden crown on His head and a sharp sickle in His hand.
>
> —Revelation 14:14

This group of four constellations adds to the promise in Virgo concerning the coming of a Son, through a virgin, who will someday be a ruler. Libra reinforces the concept that the reason for His coming will be to pay someone else's price, because they have been weighed in the balance and found wanting. Bootés and Crux, which are the last decans of Virgo and Libra, complement the idea that this Kinsman Redeemer, like Boaz of old, will come again as a King and will have the authority to rule His creation forever.

The Constellation Scorpio and its Decans

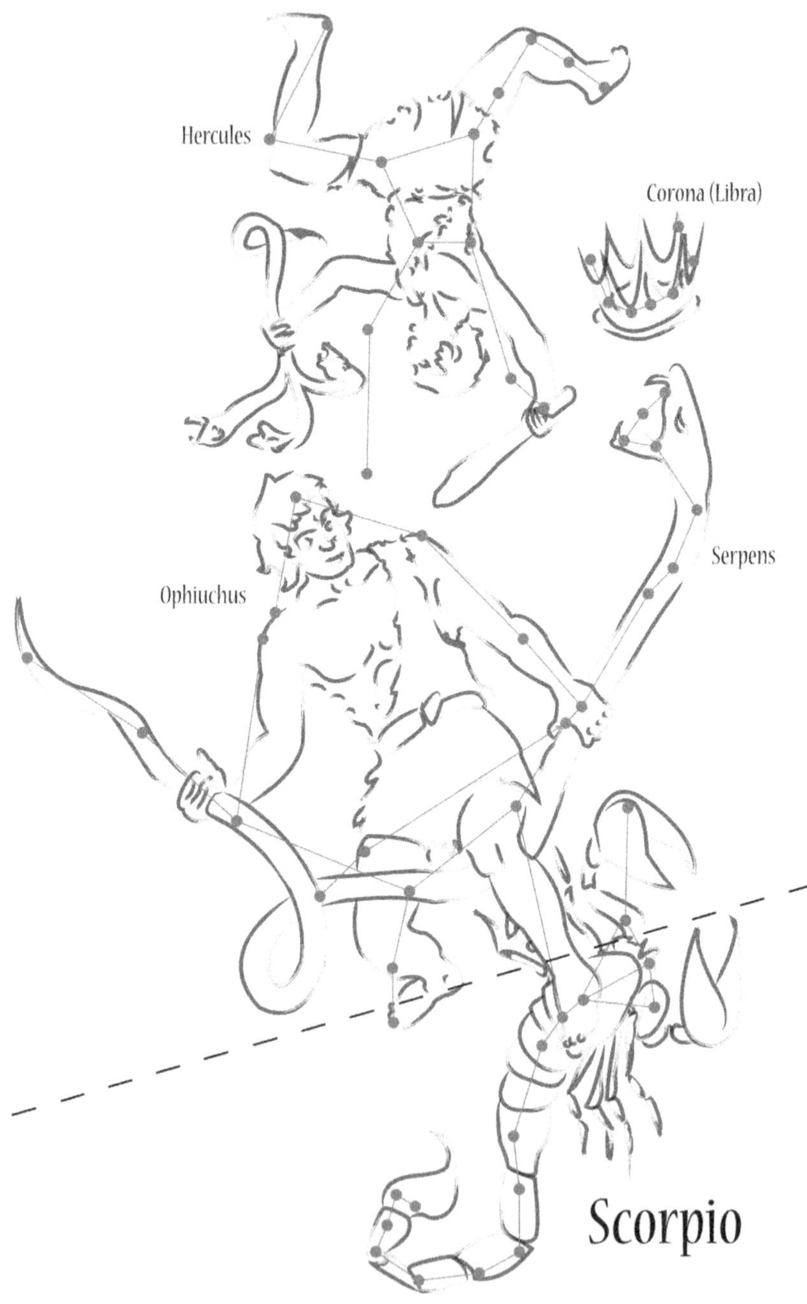

Act One, Constellation Group Three: Scorpio

Each time I describe the constellation groups — whether in writing or when teaching before an audience — I try to refrain from beginning by saying that this group tells the most profound story of all because the decans are all revealing and fascinating in their own right. However, the story this set of three decans speaks about is truly unprecedented in its clarity and its parallelism with God's plan, as revealed in His Word.

Scorpio appears in the night sky, of course, as a scorpion. It arrives overhead at midnight during the month of June, with the constellation Taurus rising with the sun six hours later. Given its three decans, these four constellations taken together suggest a great conflict.

A mighty warrior takes center stage within this group. He is struggling with a huge serpent held in his hands as he fights to get control of this intertwining beast. Meanwhile, above his head is another warrior who is also battling another snakelike creature. But the most fascinating feature in this group is what is at the feet of this first mighty warrior: Scorpio. Mankind's ultimate enemy is attempting to strike the warrior's heal, while the warrior is taking its life, at the same time, by striking it on its head with his other foot.

Does that ring any bells? Recall the battle that God said in Genesis 3 would occur between the offspring of Eve and Satan. Satan's offspring would wound Eve's offspring in the heel, while Eve's offspring would someday come and kill the offspring of Satan by striking him in the head. Two revelations, but one and the same story! Do you think that they might have come from the same source, namely God?

Satan and his minions are represented several times within this group, with Scorpio being the primary representation. With his tail, Satan is seen doing exactly what he attempted to do during the life of Yeshua on earth. Whether his efforts came after Yeshua's forty days of fasting or at the cross, Satan attempted to suppress and to deter – and if he couldn't do that – to kill the offspring of Eve, the very offspring who long ago was prophesied to come. And Satan did succeed in killing our Messiah. However, the adversary didn't realize that, in doing so, he fulfilled the very plan of God that would ultimately result in the saving of mankind, even as it also brought on his own destruction.

The Mighty Warrior

The decan *Ophiuchus* is the mighty warrior described above. He is fighting with a snake while his heel is being struck by the tail of the scorpion, and he is striking the head of the scorpion with his other heel. This snake is called *Serpens*, the next decan, which is the second representation of Satan within this group. Ophiuchus, representing the seed of Eve and the son of Virgo, is preventing the snake from reaching a crown that appears over its head. With all of its might this serpent is stretching out its head, reaching above and thereby revealing its true goal: rulership. He wants to conquer and reign with all the power of a king.

> Then I saw a beast coming up out of the sea, having
> ten horns and seven heads, and on his horns were ten
> diadems, and on his heads were blasphemous names.
> 2 And the beast which I saw was like a leopard, and his
> feet were like those of a bear, and his mouth like the
> mouth of a lion. And the dragon gave him his power
> and his throne and great authority. 3 I saw one of his
> heads as if it had been slain, and his fatal wound was
> healed. And the whole earth was amazed and followed
> after the beast; 4 they worshiped the dragon because he
> gave his authority to the beast; and they worshiped the
> beast, saying, "Who is like the beast, and who is able to
> wage war with him?" 5 There was given to him a mouth
> speaking arrogant words and blasphemies, and author-
> ity to act for forty-two months was given to him. 6 And
> he opened his mouth in blasphemies against God, to
> blaspheme His name and His tabernacle, that is, those
> who dwell in heaven.
>
> 7 It was also given to him to make war with the saints
> and to overcome them, and authority over every tribe
> and people and tongue and nation was given to him.
> 8 All who dwell on the earth will worship him, everyone
> whose name has not been written from the foundation

> of the world in the book of life of the Lamb who has been slain.
>
> —REVELATION 13:1–8

The above passage adds to this constellation's tale. For a time, the beast we know as Satan thinks that he has achieved the primary goal of his struggle throughout time — becoming the ruler of the entire world. However, just as he seats himself on the very throne of God in the above passage from the book of Revelation, and makes his fateful proclamations of godhood, the seventh trumpet blast occurs.

Revelation 11 proclaims that the forty-two months given to Satan to rule actually belong to the true Creator.

> [15] Then the seventh angel sounded; and there were loud voices in heaven, saying, "The kingdom of the world has become the kingdom of our Lord and of His Christ; and He will reign forever and ever." [16] And the twenty-four elders, who sit on their thrones before God, fell on their faces and worshiped God, [17] saying, "We give You thanks, O Lord God, the Almighty, who are and who were, because You have taken Your great power and have begun to reign. [18] And the nations were enraged, and Your wrath came, and the time came for the dead to be judged, and the time to reward Your bond-servants the prophets and the saints and those who fear Your name, the small and the great, and to destroy those who destroy the earth." [19] And the temple of God which is in heaven was opened; and the ark of His covenant appeared in His temple, and there were flashes of lightning and sounds and peals of thunder and an earthquake and a great hailstorm.
>
> —REVELATION 11:15–19

Yes, the last three and one-half years, forty-two months, are actually the beginning of the thousand-year reign. God is seen punishing all those who have opposed Him. He is cleansing His temple and dealing with Satan, the beast, and the False Prophet.

With that accomplished, God then begins His reign by showing off His bride and giving the queen her rightful place on the throne. But that's getting ahead of our story!

The third decan in this group is *Hercules*, whose Hebrew name is *Gibbor*, meaning "mighty." Hercules is holding a multi-headed beast. Sometimes this beast has three or even seven heads, mirroring the beast described above in Revelation 13 — which, of course, is Satan appearing for the third time in this group of decans. With a club in his other hand, Hercules completes his work by demolishing him.

Hercules then ends the story of this group as it began with Scorpio. The snake, the beast, and Satan all succumb to the authority and might of the true King. However, in the process there will be a mighty battle. Those without hope might wonder who will be the victor in the end. But those who have had hope, throughout the ages, will not be disappointed. God absolutely wins.

So, our story has now developed further. A virgin will come, bearing a son. This son will be in conflict with much of mankind and will ultimately die on a cross. The reason for His death is revealed in the second group of constellations, where someone will be weighed and found wanting. A price must be paid. The one who pays that price is the son, but He will ultimately be the King. This King will enter into a great battle with a mighty foe: a huge snake who will lust after the King's crown. The snake will desire royalty and a kingdom but will fall short of accomplishing its goal.

The Center of our Constellations Has Not Always Been Polaris

The Bible reveals that the ultimate source of all this evil is not to be sought in Nimrod the man (the first of the post-Flood antichrists), but rather in the evil character of the one who possessed him — namely, Satan. In many passages in the Bible the following associations with Satan are made: the serpent in Eden, Leviathan the sea monster, the dragon, "that ancient serpent," "the god of this age," the king of Babylon, the king of Tyre (Phoenicia), the pharaoh of Egypt, the father of lies, the prince of the power of the air, and others as well.

In Isaiah 14 he is spoken of as the instigator of war in the heavens for attempting to "ascend to the heavens" in order to raise his throne "above the stars of God," and to sit "on the mount of assembly in the recesses of the north" and thus to rule the universe. Satan is therefore the "lord of (the) rebellion" and "lord of the black void of the north."

Primeval astronomy, of which Babylonian astrology (still extant today) was a corruption, was based on the realization that the entire universe was created and had worth only in relation to the earth. Thus the ancients saw it as no accident that the stars and planets were set in a certain order by God. (See the classic books by Bullinger and Seiss on this subject listed in the bibliography at the end of this book.) The antediluvian patriarchs developed a system of constellations to serve as perpetual reminders of man's fall and the promise of a coming redeemer, as well as a record of the angelic conflict down through the ages.

At the most prominent place in the heavens the patriarchs placed the constellation Draco, the dragon, which lies coiled about that point of the sky they called "absolute north." This was the center of the circle that the earth's north pole describes in the sky every 25,858 years. About 4000 BC, the star Iota Draconis was the visible star nearest to the north pole, while about 3000 BC the north pole centered exactly on the star Alpha Draconis. This one was also called Thuban and is the brightest star in the constellation.

This portion of the dragon is depicted as attempting to encoil the constellation Ursa Minor, which was originally called the "little flock" or "little sheepfold" because it represented the faithful remnant of Israel, the people of God. We find this exact picture in Revelation 12, where the text describes events yet to be enacted in human history. The most devastating battle of all time is yet to be fought on earth and in space (i.e., "the heavens"). The pole star today is, of course, Polaris in Ursa Minor, and it will next enter the constellation Cepheus, which pictures God as the triumphant King over all the earth.

It is also notable that in primeval astronomy the dragon's head is shown as being crushed under the foot of a hero who, at the same time, is using a club to beat to death the Hydra who has stolen the fruit of immortality. Head-to-head with this hero but set in mirror image across from him is a second hero, grappling with a huge snake whose gaping jaws are straining to grasp Corona Borealis, the Crown of the North. This second hero is also crushing a vile enemy underfoot, but this time it is the scorpion. Yet even as he does this another scorpion bites his heel. This early configuration of the constellations around the north pole was derived from biblical ideas about the events recorded much later in Genesis.

The Babylonian Creation Epic describes Marduk leading a rebellion of the gods against Tiamat, who has planned destruction for them. The Hebrew cognate for Tiamat is *tehom*, used in the Bible only to describe "the deep" upon which God moved at the beginning of creation. Later a part of the tehom was imprisoned within the bowels of the earth (in Jewish rabbinical tradition) and opened to release the "waters from below" at the same time the vapor canopy collapsed during the Flood, in order to destroy the civilization of Noah's day. This destruction is said to have come about because of excessive influence by Satan in the affairs of men, such as intermarriage with mortals producing giants on the earth with various genetic defects of a serious nature. In the Babylonian version Marduk wins and is eulogized by the other gods in a list of fifty names to which can be traced most of the gods of antiquity. This epic was read aloud every New Year's day in Babylon in front of the statue of Marduk.[16]

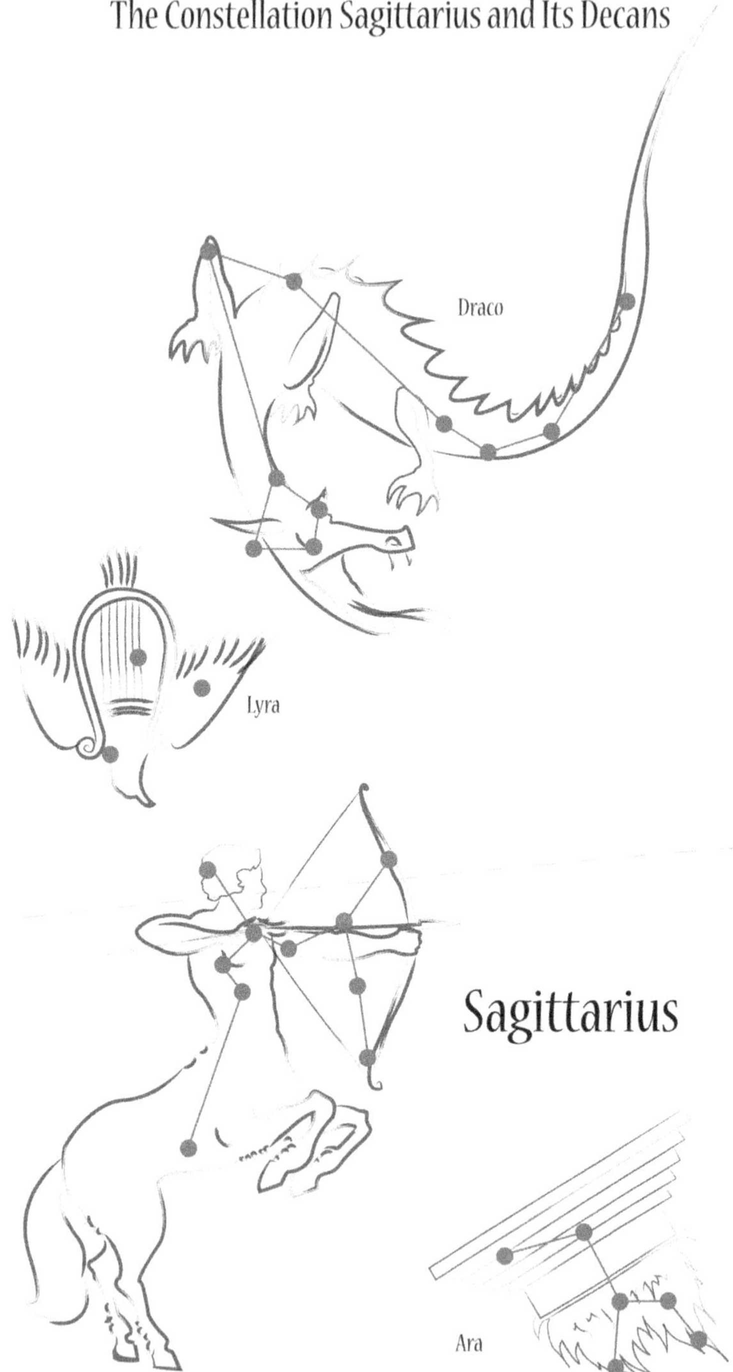
The Constellation Sagittarius and Its Decans
Draco
Lyra
Sagittarius
Ara

Act One, Constellation Group Four: Sagittarius

With this constellation, we come to the conclusion of Act One. It is fitting that the last scene of Act One closes with constellations and decans that speak about the two main players seen in the previous three scenes. *Draco* is one of the decans in this group. He represents the scorpion, the snake, and the dragon of the previous constellations whom we have learned represent Satan. And Sagittarius, the primary constellation in this group represents the Son, Centaur, Hercules, and Lamb decans.

We have met these constellations in the previous scenes of Act One. We now know that they represent our Creator who will be King. These two enemies have one more act to play out, but this time both of their destinies are uncovered. Let's now explore what this last constellation unveils about the results of all of the previous conflicts, battles, and deaths.

Sagittarius is the principle constellation in Group Four. This constellation is very similar to Centaurus, one of the decans found in Virgo. Both appear as centaurs; that is, as horses with the torso and head of a man, thus revealing their multiple natures. Both also have weapons. Centaurus has a spear and a shield; Sagittarius has a bow and a quiver full of arrows.

The first is aiming at the Victim, a lamb-like animal. The second is aiming at the heart of the scorpion we met in the previous constellation group. It seems that the Lamb must die in order for it to have the authority that it needs to deal with the second target. This authority, of course is not required by God to do away with Satan. But by His death He does obtain the authority over sin, by paying our price for our sin. We are the ones found wanting by the scales of Libra.

> "The Spirit of the Lord is on me, because he has anointed me to proclaim good news to the poor. He has sent me to proclaim freedom for the prisoners and recovery of sight for the blind, to set the oppressed free."
>
> —Luke 4:18, NIV

With that newfound authority Sagittarius is able to set the captives free. Our sin was really the Lamb's first target and opponent. By literally becoming sin and thus temporarily distancing Himself from God the Father, Yeshua was moved to cry out on the cross:

> "My God, my God, why have you forsaken me?"
>
> —Matthew 27:46, NIV

When He became sin for our sake He had to be abandoned by the Father, but just for a short while. His second enemy is the Deceiver himself, here pictured as Scorpio. This constellation, the fourth to appear in the procession in the sky, takes its place overhead at midnight in July. Gemini accompanies the rising of the sun this same time of year.

The first decan is *Lyra*, or the "Eagle." The ancient maps of the stars show both an eagle and a lyre, or harp. Its brightest star is known as Vega, which means "He shall be exalted." The harp is used in Scripture as an instrument to worship and bring praises to God.

> Give thanks to the LORD with the lyre; sing praises to Him with a harp of ten strings.
>
> —Psalm 33:2

> [2] And I saw what looked like a sea of glass glowing with fire and, standing beside the sea, those who had been victorious over the beast and its image and over the number of its name. They held harps given them by God [3] and sang the song of God's servant Moses and of the Lamb:
>
> "Great and marvelous are your deeds, Lord God Almighty. Just and true are your ways, King of the nations. [4] Who will not fear you, Lord, and bring glory to your name? For you alone are holy. All nations will come and worship before you, for your righteous acts have been revealed."
>
> –Revelation 15:2–4

Lyra the harp is bringing praise and thanks to the Two-Natured One, the One who has paid our debt and delivered us from our bondage. God's people are seen in the words of Revelation 15, playing harps while singing the song of the Lamb and the song of Moses. Both of these songs are taken from Torah and were sung by the Israelites after the crossing of the Red Sea in Exodus 15 and Deuteronomy 32 just before Moses departed to die on Mount Nebo. These songs include praises to God, who had delivered the Israelites from their opponents and from themselves.

The second decan in this group is *Ara*, which is an altar placed upside down. The same word was used by the Greeks as a curse and suggests that this decan is the punishment for the loser of the battle described in Act One. Certainly the Word of God suggests the same: A place of fire is reserved for the devil and his angels according to Matthew 25:41. Revelation suggests the same destiny for our adversary:

> And the devil who deceived them was thrown into the lake of fire and brimstone, where the beast and the false prophet are also; and they will be tormented day and night forever and ever.
>
> —Revelation 20:10

Who Is at the Center of Your World?

Finally, we arrive at the last decan in the last scene of Act One — *Draco*. Four to five thousand years ago, Thuban, one of the stars in Draco, was the pole star. Remember, the pole star is the North Star, which all the constellations and stars revolve around from our earthly perspective. This emulates Satan's own longstanding desire. He very much wants mankind to focus and spin around him. Satan wants to be at the center of God's creation, being worshiped by all and having the authority to rule over everything.

Over time, as the story in the sky reveals, the center has migrated due to the procession of the stars, away from Draco. Instead of the stars of Draco being used for navigation, as the ancients did, the North Star is now Polaris. The "Guiding Star" has taken on the place of the dragon!

What light do you use for navigation in your life? The constellations of the fourth chapter of Act One expose the victor of the battles between the Son and the dragon. Praises from all are directed to the Son, the winner. Thus Draco, the dragon, is left with only one destiny, punishment in the Lake of Fire.

This theme is repeated in each of the three Acts. Act One ends with the dragon being cast out. As we will explain in the next chapters, Act Two will end with Cetus, the sea monster, being bound. And Act Three will conclude with the destruction of Hydra.

The Constellation Capricorn and Its Decans

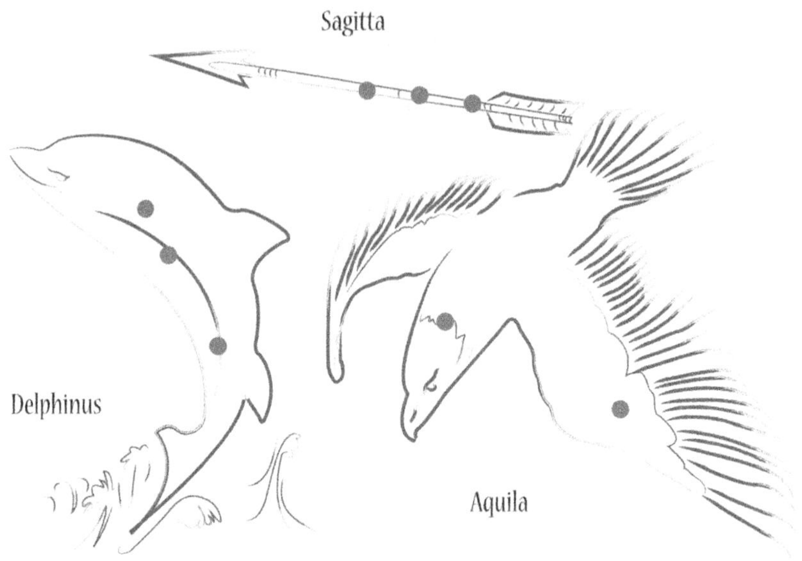

Chapter Eleven

The Story in the Sky: Act Two

Act Two, Constellation Group One: Capricorn

The previous act focused on the conflict between the Son and the dragon. Very little attention was given to why the battle had to occur in the first place. Were the two just fighting with each other because their natures were at extreme odds? Or, did something else drive them to this combat to the death?

There are hints in the first act, such as the scales of Libra. But we are not informed about exactly who or what is being weighed, and who or what is found falling short. All we know is that this Son is determined and willing to come and bring balance back to the scales. Surprisingly, the cost to accomplish this feat will be His life. The obvious question from Act One, then, is what is so important and valuable to the Son that He would be willing to pay such a price?

The answer to this question is the focus of Act Two. In fact, the very first constellation, and the primary group of stars in Group One, begin to reveal the answer in this Act. Capricorn pictures the front end of a goat and the back end of a fish. Goats were commonly used, especially in the Bible, as sacrifices offered up to God for the forgiveness of sin.

This constellation, is the fifth to appear in the procession in the sky in the course of the year. It appears overhead in the month of August. The constellation Cancer accompanies the rising of the sun at this same time of the year.

How God Wishes to Be Approached

At both the temple in Jerusalem and the tabernacle in the wilderness, the throat of each sacrificial animal would be cut before it would be placed on the bronze altar in the outer court of the temple grounds. This offering was called *korban* in Hebrew. The root word that korban comes from is *karav*. Its primary meaning is "to approach or draw near."

In other words, within the word for sacrifice God embedded His primary message to mankind. As you humble yourself by recognizing your shortfalls you may draw near to Me. This is why God tells the Israelites that He hates their sacrifices.

> Alas, you who are longing for the day of the LORD, for
> what purpose will the day of the LORD be to you? It will
> be darkness and not light; 19 as when a man flees from a
> lion and a bear meets him, or goes home, leans his hand
> against the wall and a snake bites him. 20 Will not the
> day of the LORD be darkness instead of light, even gloom
> with no brightness in it?
>
> 21 "I hate, I reject your festivals, nor do I delight in your
> solemn assemblies. 22 Even though you offer up to Me
> burnt offerings and your grain offerings, I will not accept
> them; and I will not even look at the peace offerings of
> your fatlings. 23 Take away from Me the noise of your
> songs; I will not even listen to the sound of your harps.
> 24 But let justice roll down like waters and righteousness
> like an ever-flowing stream."
>
> —AMOS 5:18–24

He is not rescinding His commands with respect to the offering of sacrifices. Rather, Elohim is telling us that offering up the symbol without real repentance is just the empty practice of religion rather than its true, meaningful substance. The Hebrew word for religion is *dot*. The pictographic message embedded within the letter symbols communicates to us what true religion is. The pictographic understanding of *dalet, tav* tells us that our religious practices should be our pathway to the sign of the covenant. If they do not reward us with greater intimacy with our Creator we are defeating their purpose.

> Pure and undefiled religion in the sight of our God and Father is this: to visit orphans and widows in their distress, and to keep oneself unstained by the world.
>
> —JAMES 1:27

The sacrifice of animals was to remind the Israelites of the Son who would come as a perfect Lamb and would lay down His life willingly, as an offering to pay the price for sin. This had been taught through many passages in Torah, given to the Israelites on Mt. Sinai. The messages in the sky would have given them a further foundation for understanding.

As man recalls the price that was paid for his sin, God's hope is that we will approach Him humbly, ask for His forgiveness, and repent.

Rebirth and Our True Nature

The Hebrew word for repent is *shuv*, spelled *shin, vav, vet*. Its primary meaning is "to turn about, to return." The idea is that, if one sins, true repentance is not just asking for forgiveness but also includes turning away from our sinful ways and returning to our true nature.

This true nature derives from our rebirth as sons and daughters of God, for at that moment we are no longer sons of man, characterized by the sinful nature we acquired from the first Adam. We now carry the nature of our new Father, which is righteousness. And, we are now called to a new walk, one of holiness. A walk of holiness is the way a righteous son or daughter acts and thinks.

> "For I am the Lord who brought you up from the land of Egypt to be your God; thus you shall be holy, for I am holy.'"
>
> —Leviticus 11:45

> "Speak to all the congregation of the sons of Israel and say to them, 'You shall be holy, for I the Lord your God am holy.'"
>
> —Leviticus 19:2

> Because it is written, "You shall be holy, for I am holy."
>
> —I Peter 1:16

God wants us to double our efforts to draw close to Him by trying to understand His nature. He wants us to be friends, then sons and daughters, and ultimately His bride.

God's ultimate goal is revealed by Capricorn. As a man and a woman become one when they become married, we become one with our Creator through the covenants of service, friendship, and inheritance, finally becoming His bride.

The goat represents the perfect sacrifice, the Son of God, and the fish represents mankind. In Scripture, fish are sometimes used metaphorically to represent people, or Israel.

> And He said to them, "Follow Me, and I will make you fishers of men."
>
> —Matthew 4:19

> "Again, the kingdom of heaven is like a dragnet cast into the sea, and gathering fish of every kind; [48] and when it was filled, they drew it up on the beach; and they sat down and gathered the good fish into containers, but the bad they threw away."
>
> —Matthew 13:47–48

> "'As the Lord lives, who brought up the sons of Israel from the land of the north and from all the countries

> where He had banished them.' For I will restore them to their own land which I gave to their fathers. 16 Behold, I am going to send for many fishermen," declares the Lord, "and they will fish for them; and afterwards I will send for many hunters, and they will hunt them from every mountain and every hill and from the clefts of the rocks."
>
> —JEREMIAH 16:15–16
>
> "It will come about that every living creature which swarms in every place where the river goes, will live. And there will be very many fish, for these waters go there and the others become fresh; so everything will live where the river goes."
>
> —EZEKIEL 47:9

In the Hebrew aleph-bet, the letter which represents life is *nun*, a picture of a fish darting through water.

God's desire is to reunite and restore the broken relationship between us and Himself. Someday He will accomplish His goal and place His bride on a throne at His side, from which they will rule and reign together over creation.

The Arrow of the Almighty

Sagitta, the "arrow," is the first decan in Act Two, Constellation Group One, of which Capricorn is the primary constellation. The Hebrew name for this decan is *Sham*. Many students in this field claim that this word means "destroying." On that definition they then base various speculations about what this arrow might truly mean.

The problem with all of this is that most of these commentators use a great deal of material from E. W. Bullinger. He was the original researcher to discover the constellations' biblical revelations. Much of his work was profound, but on this point I cannot verify his research. The word *shamad* in Hebrew means "to destroy" or "destruction." This word is spelled *shin, mem, dalet*. The word *shamah* means "wasting" or "desolation" and is spelled *shen, mem, hey*.[17]

- It is possible that Bullinger used these words as a source for his definition of this decan's Hebrew name. In Hebrew, *Sham* would be spelled *shin, mem*. This spelling leads to the more common pronunciation of "shem." This word is more widely known because it was the name of one of the sons of Noah who accompanied him in the ark.

Shem or Melchizedek

Among Hebrew scholars the same Shem (son of Noah) is known later in the biblical text as Melchizedek, the king of Salem.

> And Melchizedek king of Salem brought out bread and wine; now he was a priest of God Most High.
>
> GENESIS 14:18

Jerusalem and Salem are the same city. The word *Jerusalem* is made up of two Hebrew words. *Salem* is better known in its other pronunciation, "shalom," which means "peace." It is commonly used by Jews today to greet someone. It means "hello," or "peace be to you."

Jeru, the first part of Jerusalem, is thought to mean "men" or "people," or perhaps "foundation" or "habitation."[18] It is appropriate, then, that Jerusalem will be the dwelling place of the coming King, who will truly be the foundation for peace on earth for all mankind.

However, there are differing views with respect to the meaning of Jerusalem. Dr. Danny Ben-Gigi proposes the following:

> *Yir-oo-shalem* is made of two words: *yir-oo* and *shalem*. The root of the word *yi-r-oo* means, "They will see," or "They will feel awe." "Shalem" means "complete, whole." "They will see completeness," or "feel the awe of wholeness" is the meaning of the name Jerusalem.[19]

Maybe this is why everyone is fighting over this city. Doesn't everyone want to be restored back to wholeness, to recover from the consequences of sin?

Less commonly known is that the word *shem* means "name." In fact, the phrase *Ha Shem* is used often in the Hebrew language to refer to God (as in "The Name'). In this Semitic language all the letters are consonants. The vowel sounds are indicated by vowel points, little markings in the form of dots or dashes above, to the side, or below the consonants. Longer words are formed around root words. So, if you know the consonants in a word, at the very least you can know the root meaning of that particular group of consonants.

This is not so in English. The word "good" is not associated with "goad" even though their consonants are exactly the same. In Hebrew these sounds would have similar root meanings.

Given all this, the decan in the constellation Capricorn is sending us a message about God's plan. This arrow is from Ha Shem, God Himself. Its target is His Son.

> "For the arrows of the Almighty are within me, their poison my spirit drinks; the terrors of God are arrayed against me."
>
> —Job 6:4

> Then the LORD will appear over them, and His arrow will go forth like lightning; and the Lord GOD will blow the trumpet, and will march in the storm winds of the south.
>
> —Zechariah 9:14

The above verses refer to the arrows of God being used to strike down His enemies. But why is the target in this case His Son? Maybe this is the reason for Yeshua the Messiah's last statement on the cross:

> At the ninth hour Jesus cried out with a loud voice, "ELOI, ELOI, LAMA SABACHTHANI?" which is translated, "MY GOD, MY GOD, WHY HAVE YOU FORSAKEN ME?"
>
> —Mark 15:34

God allowed Himself to suffer death on the cross. Even so, this arrow came from His very own bow. He could have saved Himself. In fact, that was His cry before the events that led up to His death.

> And He went a little beyond *them,* and fell on His face and prayed, saying, "My Father, if it is possible, let this cup pass from Me; yet not as I will, but as You will."
>
> —Matthew 26:39

Some suggest that Satan won the day, for this arrow must be from the adversary. They suppose that things would be so much better if our Lord had not been killed. They then play the blame game: "It's the Jews' fault for the death of our Lord!" Thus they find comfort and excuses for allowing or even perpetrating pogroms — persecutions and crimes against the descendants of Abraham. In justifying their crimes they often refer to Jews as "Christ killers," which was the name the Nazis used as they led millions of Jews to their deaths in the concentration camps.

However, as I mentioned earlier, Jews did not possess the authority to sentence someone to death two thousand years ago. Only the Romans could originate and execute this type of sentence, and that is exactly what they did. The truth is, God allowed His own life to be taken at the hands of the very ones He came to save: mankind, meaning ALL of mankind.

It wasn't the Jews or the Gentiles or Satan, although our adversary was probably salivating over the idea. At the time he thought he'd won a great victory. What he didn't expect was that Yeshua would follow His death with the greatest recovery the world has ever seen. Of course, power over death has been the desire of man for all time, even though we have failed to achieve it at every turn. The Fountain of Life can only flow forth from the One who has paid the price for our death sentence.

Certainly, God has used people and Satan throughout the ages to get His job done. And Satan has taken advantage of us as well. But God has promised that, if we love Him, all things will work out for our good even if they were intended for evil, such as the evil deeds of our adversary.

> And we know that God causes all things to work together for good to those who love God, to those who are called according to His purpose.
>
> —Romans 8:28

Satan, still thinking that he is in control, is sometimes referred to in the Bible as a bowman. Revelation 6:1 describes a white horse with a rider holding a bow. This archer has already shot his arrow, for he is described as having just a bow. We have learned in the *Lost in Translation* book series that Satan has copied and will continue to copy God until his very end.

He does this to deceive mankind into thinking that he is the real Creator. Satan will claim to be raised from the dead, and thus to have power over death. He will be accompanied by a prophet who will claim to be Elijah, just like the prophets of old foretold the second coming of God. He will offer power and rewards, and he will have a bride just like the real Groom. And in Revelation 6 he will come riding on a white horse, wearing a crown, claiming to be king and deceiving almost everyone in the process by copying the real King. On the contrary, Revelation 19 describes the real God, who finally does come at the end of the tribulation period, to set up the kingdom that will last forever.

The decan Sagitta might very well be the missing arrow from the bow of the archer above, but his purposes will work out for good just like all the other plans of our Lord.

Wounded and Falling

The second decan in the constellation Capricorn is *Aquila*. This eagle is always depicted as falling out of the sky. "The Wounding" is the name of one of its stars, confirming the idea that this bird of prey has been wounded in a great fall. Certainly this scene copies the wounding Yeshua Himself incurred on our behalf on the cross. Perhaps this group of decans is suggesting that the arrow that was shot from the bow of Satan was meant for mankind. But in the nick of time God flew into its path and intercepted it before it reached our hearts. This arrow was meant for evil but God used it for good to advance His own purposes.

God is described many times in the Scriptures as an eagle. In Exodus 19:4 and again in Revelation 12:14, an eagle swooping in to defend, protect, and remove God's people from harm and take them to a safe place is used to deliver a promise to His people. Aquila has fulfilled this same promise, this time preventing the arrow from accomplishing its goal. The enmity and hatred Satan has for God's people certainly was helping the arrow on its way. But the love God has for His own is far greater.

From the Sea of Humanity

A dolphin, full of life, springing out the sea, is the last decan in this group. *Delphinus* is the perfect picture of our Savior. A body of water or a sea is used in Scripture to refer to mankind, as in Revelation 15. Here Yeshua, who came as a man, is able to rise out of the sea of humanity with His victory over sin. He rose from the dead early on the first day of the week, which was also the feast of Firstfruits. This dolphin represents the same promise to mankind: "As I now have the power over death, I will come again and give you everlasting life as well."

However, this promise carries with it a condition foretold by the two preceding feasts. Passover and Unleavened Bread tell of a Savior who would come and pay the price for our sin. Our responsibility would then be to believe that this payment was needed, and to trust in Him to accomplish His promise to deliver us from our debt. This faith, God says, will be confirmed by our response to His presence in our lives.

> [20] You foolish person, do you want evidence that faith without deeds is useless? [21] Was not our father Abraham considered righteous for what he did when he offered his son Isaac on the altar? [22] You see that his faith and his actions were working together, and his faith was made complete by what he did. [23] And the scripture was fulfilled that says, "Abraham believed God, and it was

> credited to him as righteousness," and he was called God's friend.
>
> —JAMES 2:20–23, NIV

Delphinus the dolphin is a picture of mankind as well, springing back to life via Yeshua's victory over death. Taken together, these groups of star pictures tell of a oneness that will come between God and mankind. This union, however, will come at great cost. The Eagle will stand up before the arrow and take its death stroke at full force. Yet God will be (and always has been!) able to rise above death and offer the same gift to all.

The Constellation Aquarius and its Decans

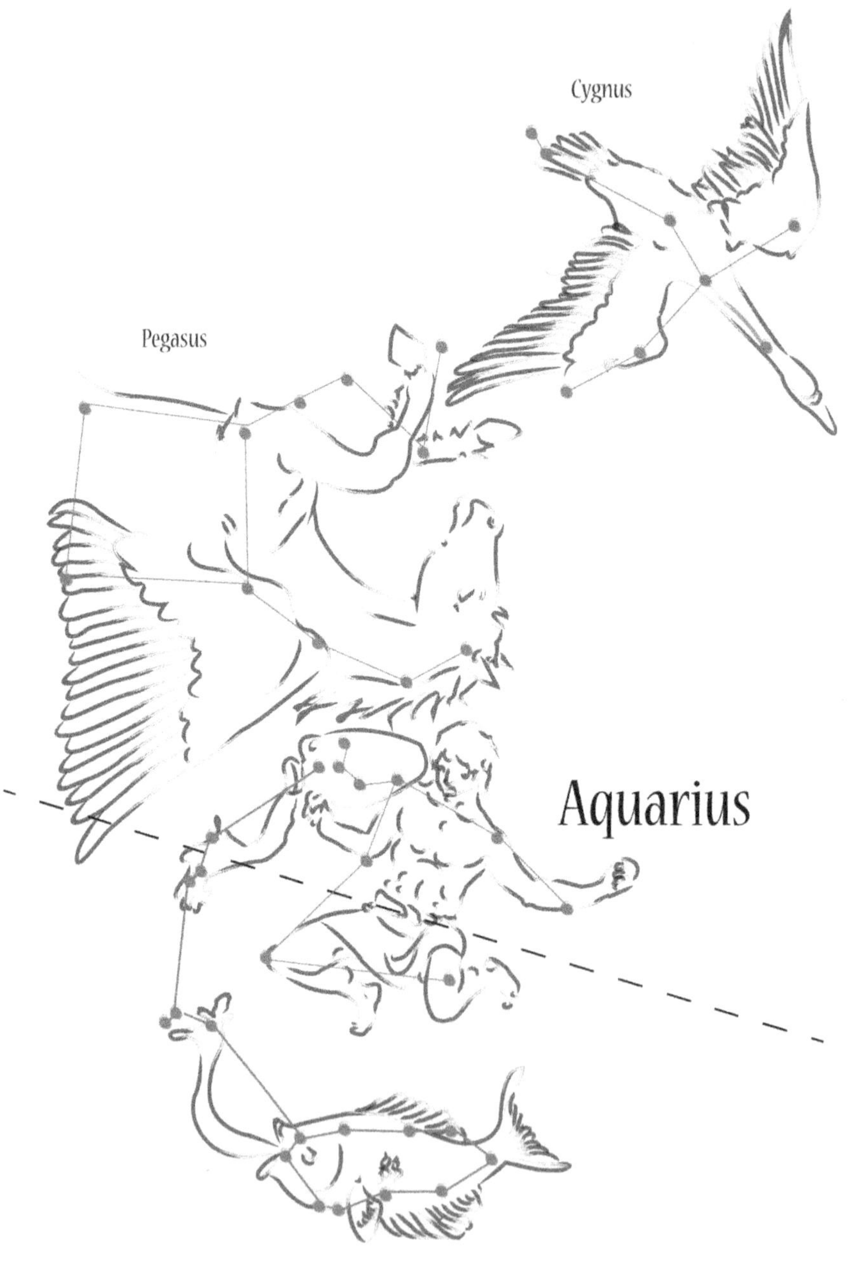

Act Two, Constellation Group Two: Aquarius

The constellation Aquarius, the sixth to appear in the procession in the sky, takes its place overhead at midnight in September. The Mighty Leo accompanies the rising of the sun during this same month. Act One only hinted at the purpose for the coming of this child through the virgin, Virgo. Act Two, starting with Capricorn, begins to reveal the Son's purpose. As a result, the ancients start to realize that they themselves were, in fact, the reason for the death of the child.

Aquarius builds on this theme, again using a fish and a stream of lifesaving water flowing forth from an urn held by this man-child. This time, though, the child has grown up. The image suggests that the fish was in much need of salvation, for it swims at the end of this stream, gobbling up the entire flow.

> "But whoever drinks of the water that I will give him shall never thirst; but the water that I will give him will become in him a well of water springing up to eternal life."
>
> —John 4:14

> Jesus said to them, "I am the bread of life; he who comes to Me will not hunger, and he who believes in Me will never thirst."
>
> - John 6.35

These verses suggest that the Son did indeed come with a purpose, which was to offer to mankind eternal life. This everlasting life is represented as water that satisfied our thirst.

The biblical linkage between flowing water and life is not confined to the book of John. From the very beginning, Genesis describes a river that flowed eastward from the presence of God to the Garden of Eden. There it split into four rivers as it distributed its life-giving flow to the plants and animals living in the garden. Possibly this same water flow is described in Revelation 22.

> Then he showed me a river of the water of life, clear as crystal, coming from the throne of God and of the Lamb,
> 2 in the middle of its street. On either side of the river

> was the tree of life, bearing twelve kinds of fruit, yielding its fruit every month; and the leaves of the tree were for the healing of the nations.
>
> —Revelation 22:1, 2

Here this river is described as having the water of life. It comes from the presence of God. Of course, God will take back His dwelling place from the False Messiah and set up His kingdom at the temple in Jerusalem.

There is a river today that flows from the temple area through the Kidron Valley. Maybe a better description would be a small intermittent stream, which is also known to scientists as an ephemeral flow. This seasonal flow works its way eastward, eventually leading any water in it to the Dead Sea. Someday, this flow will be greatly increased and will water a new garden along its banks.

This river and its greatly increased flow is further described in Ezekiel 47. The river in this passage starts out small, just like today. But as God ushers in the final restoration of His creation, this flow will increase, finally becoming a large river. The waters will begin their course at the throne of God in Jerusalem and will nourish a great variety of trees as the waters work their way eastward. Their effect on the Dead Sea will be startling. Once again it will come to life; fish will abound, feeding all who come there.

> "How fair are your tents, O Jacob, Your dwellings, O Israel! [6] Like valleys that stretch out, like gardens beside the river, like aloes planted by the LORD, like cedars beside the waters. [8] God brings him out of Egypt, He is for him like the horns of the wild ox. He will devour the nations who are his adversaries, and will crush their bones in pieces, and shatter them with his arrows. [9] He couches, he lies down as a lion, and as a lion, who dares rouse him? Blessed is everyone who blesses you, and cursed is everyone who curses you."
>
> —Numbers 24:5, 6, 8, 9

This river of life is exactly what God has promised to Israel. In the prophecies in Numbers 24, which are being fulfilled today, Yahweh will cause the land of Israel to flourish. The whole world has opposed Israel at almost every turn, but it has kept its place in the land, becoming a leading exporter of cutting-edge technology. Israel has also turned what was once a dustbowl of a country into a lush landscape, which produces bountiful harvests.

The prophecy in Numbers also has a message for those who would oppose Israel: God will come against you. It is not accidental that, since the creation of the state of Israel, many nations have attempted to eliminate this new state through political, financial, and military means. All have failed in their attempts and have become weaker themselves as a result of all their destructive efforts.

Baptism vs. Mikveh

In conjunction with the shopworn old premise that the New Testament presents *new* concepts and themes and support for a *new* religion, baptism is thought of by many as a "Christian" concept unrelated to anything in the Old Testament. However, as we are finding, the Bible is actually one continuous presentation of a godly lifestyle, leading to holiness.

Because, as the Bible reveals, God's overall plan for us spans the entire life cycle of mankind on earth, giving us a panoramic view of time and history. This history includes the part yet in our future, giving us hope that our destinies do not just include death, but also life after death and the coming of our Creator.

Contrary to what many Christians have been taught — and in total synch with the summary above — baptism is a very old idea with deep roots in the Old Testament. One of the instructions given to Moses when he was setting up the tabernacle included what many might call a "wash basin." This bronze laver was set between the bronze altar and the Tent of Meeting. Before anyone could enter the Inner court of the Tent of Meeting they had to wash themselves. This duty was required upon pain of death.

Thus, before a priest could approach the presence of God he needed to analyze himself, inwardly and outwardly, then purify himself via the cleansing qualities of water. God presents the idea that we are all priests, serving Him in our daily lives. The parallel is stark. As we approach God we also should present ourselves to Him as holy, humble, and repentant.

Ritual purity is described as full immersion in water. The word *mikveh*, in Hebrew, is used to describe a place where water flows together. The Hebrew baptism/mikveh was done in fresh, flowing water with the person facing the source of the flow. He would then bow forward, completely immersing himself in the moving water. The idea here is that God is the source of our life, which the flowing water represents. As we humble ourselves in front of Him, the life-giving purity washes away the impurities present in our lives due to our poor choices.

Of course, all of these rituals should only be a representation of what is going on inside our minds and hearts. Without our heartfelt commitments and decisions to act properly, all rote rituals are just water over the dam.

Aquarius and his life-giving water jug is a picture of the restoration of a river that once watered and brought life to a garden, which supported the lives of many. This river has once again been reestablished and is bringing new life back to God's people. At the time of Yeshua's return this intermittent stream will increase in size, eventually overflowing its banks and bringing blessings, life, and rebirth to a land and its people.

These people, with representatives from all other nations, will be united with their king, so well pictured by the constellation of Capricorn. This union will be possible because of the work done by the Son of Virgo. Aquarius reveals clearly this ancient purpose and plan. Someday this Son will come, after confronting the trials of the Scorpion and the cross, and will offer a lifesaving drink to all of mankind.

Remember the plan of God is to restore all things. This includes you and me, and His original creation. He may very well intend to restore it back to its original condition. After all, it was "very good."

Naming Rivers and Cities

Because one of the rivers that coursed through the original Garden of Eden was called the Euphrates, some suppose that the Garden of Eden was in the Mesopotamian valley. Today a large river by that name does indeed flow through that valley. But as is common with man, as he migrated all over the earth he also took with him the names of rivers, mountains, and even people that he was familiar with.

This phenomenon can be seen in the naming of cities, states, and even whole regions such as New York, Moscow (in Tennessee), and New England. These names were not given because the folks thought that they were in York, England or Moscow, Russia, at the time, but because of the fondness they had for the names of things back home.

This same situation may be applicable to the Euphrates. After the Flood of Noah's time, people migrated southwest from the Mt. Ararat region and came upon a great river. In memory of their original place of origin they may have named this new river after the one in the Garden. The lost Euphrates — the one that watered the Garden of Eden — flowed eastward from the presence of God. Assuming that God has not changed the location of the place He chooses to dwell in, Jerusalem is, was, and will be the center of His kingdom. East of Jerusalem, which of course is the Dead Sea's location today, could very well have been the general location of the original Garden of Eden.

And incidentally, it should not be surprising that, because Jerusalem is special in this way, there is such an ongoing fight over this city, with Satan leading the charge.

Gulping All the Water

Pisces Austrinus, the fish gulping up all the water from the jug in the arms of Aquarius, is the first decan in this constellation. Its mouth is wide open as if it is in desperate need. And so it is. As we come into faith we realize just how much we are in need of these waters of life and restoration, for very few fail to realize the terrible state our human natures are in. And when we see the One who can repair that fallen state, like the fish, we swim as fast as we can toward Him.

Or, at least, that is God's hope. Sadly, most will turn Him down to their eternal loss.

Pegasus is the second decan. The star names of this decan clearly speak of the plan of God. Such names as Returning from Afar, Who Carries, The Branch, Who Causes to Overflow, tell about the ancient story of Yeshua, who is referred to in the biblical text as *The Branch* (Isaiah 11:1, Jeremiah 23:5, Jeremiah 33:15). This Branch will return to His creation, carried on a white horse, and will bring with Him the living waters that restore life.

The discussion above makes obvious God's plans for restoration of the earth and the people who dwell there. He will return on a white horse, bringing all of this restoration with Him. We find this message in Revelation 19:

> And I saw heaven opened, and behold, a white horse, and He who sat on it *is* called Faithful and True, and in righteousness He judges and wages war.
>
> —Revelation 19:11

Could it just be coincidental that all of these skyward revelations and details just happen to line up according to the prophecies of the biblical text? As we continue to discover these parallels, the answer to that question will become even more clear.

Cygnus is the third decan in the constellation Aquarius. It takes the form of a goose with outstretched wings, flying toward its final destiny. The star names give away the messages in the stars that constitute its form: The Lord Comes, Circling and Returning, He Shall Come Down, Who Returns Quickly, and Shining Forth.

> I am coming quickly; hold fast what you have, so that no one will take your crown.
>
> —Revelation 3:11

Revelation repeats this message four times, in 22:7, 22:12, and 22:20, all of which complement the message from the star Who Returns Quickly. God's Word contains a clear message of hope concerning His return, and this message parallels Cygnus's message. For God truly desires to put an end to this age and return for His beloved.

These groups of constellations reveal the very heart of God. His desire is to come back and restore life to His people. And He will return with all the excitement of a groom coming for his bride, like a knight in shining armor saving the day.

The Constellation Pisces and Its Decans

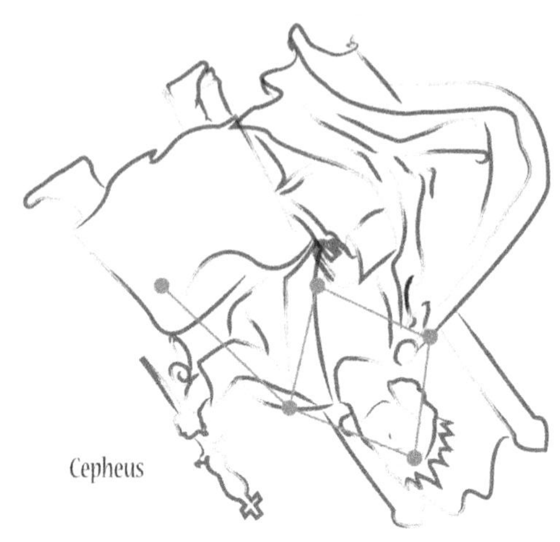

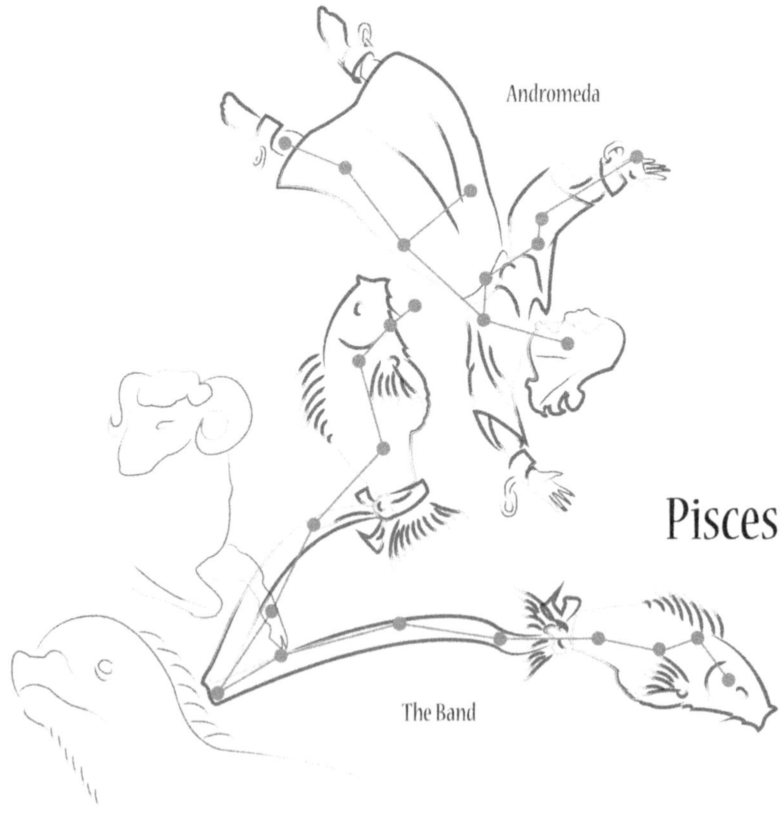

Act Two, Constellation Group Three: Pisces

The meaning of Pisces is revealed by the decan that is part of this primary constellation picture. This same association is seen in the previous constellation, Aquarius, and its first decan, Pisces Austrinus. There, a man is holding an urn from which flows a stream of water that is completely consumed by a fish.

In this constellation, Pisces is a picture of two fish attached to each other by a band. The fish are connected — one at each end — by a band, or rope. This rope, holding the fish firm, is the first decan of Group Three and (logically enough!) is called *The Band*.

The Band decan is also connected with Cetus, a decan of another constellation group. Cetus is a great sea monster and has the middle of the band held fast, and thus controls the fish connected to the band as well. Aries, the next constellation we will study, is then seen intervening into this tangle. The lamb, Aries, has interrupted Cetus's plan by extending his leg, hoof, and heel and is preventing Cetus from capturing and hauling off the two fish.

By now, all of these animal symbols are beginning to communicate a consistent message. Again and again they tell the same story, each time adding additional information and ultimately completely revealing the same basic promise. Pisces represents mankind as two fishes, which are both captured and bound. They are being led away by our foe, Satan himself, this time represented as the evil sea monster, Cetus.

Of course we now know the rest of the story. Satan does not win the day. And we have Yeshua, the Lamb, represented by the constellation Aries, to thank for that. He intervenes on our behalf and saves us from the shortcomings that Libra, the scales, exposed. These shortcomings, our sins, are the bands holding us fast to our rightful destiny of death, except for the price paid by the Lamb.

Halfway Around the Circle

The constellation Pisces (not to be confused with the first decan of Aquarius, Pisces Austrinus), is the seventh to appear in the sky. It takes its place overhead at midnight in the month of October. Virgo the virgin accompanies the rising of the sun during this same month.

At this point we have traveled exactly half of the circuit that the sun makes in the course of one year, and Virgo is now rising with the sun instead of being overhead at midnight. This new "opposite" orientation can be seen by noticing that, as Virgo is rising with the sun, Pisces is setting at the same time during this same time of year.

Some suggest that the two fishes of Pisces that are being held fast by the decan The Band represent Judah and Ephraim, the two ancient nations of Israel. Recall that Judah represented the tribes of Benjamin and Judah along with many of the Levites, and the northern ten tribes represented the remaining sons of Israel. These two groups split into two nations following the conflict that occurred after the death of King Solomon.

Ezekiel 37 tells us that someday the ten northern tribes and the two southern tribes will be reunited. Some suggest that these two fish are seen here, in this cluster of constellations, as being brought together and therefore fulfilling Ezekiel's words.

However, these two fish pictured in the constellation Pisces represent a much larger catch. God does want His people back, represented by the sons and daughters of Israel. However, the concept of Israel represents more than just the people who have a bloodline back to Jacob, whose name was changed by God to *Israel*. As Gentiles become believers in the Messiah of God's people, they are also grafted into the family of God known as Israel. When speaking to the Gentiles at Rome, Paul taught the following:

> [17] But if some of the branches were broken off, and you, being a wild olive, were grafted in among them and became partaker with them of the rich root of the olive tree, [18] do not be arrogant toward the branches.
>
> —Romans 11:17, 18

> For if you were cut off from what is by nature a wild olive tree, and were grafted contrary to nature into a

> cultivated olive tree, how much more will these who are the natural *branches* be grafted into their own olive tree?
> —Romans 11:24

According to Paul, the descendants of Jacob/Israel were represented by a cultivated olive tree. Some of its branches were cut off because of the people's unbelief. Meanwhile, some of the Gentiles were grafted into the *cultivated* olive tree (i.e., Israel) from *wild* olive trees. The resulting tree thus represents all of mankind. In the Bible the word *Israel* really represents the bride of Yeshua, not just believing Hebrews but believing Gentiles as well. All have been invited.

And no, I am not trying to steal Jewish identity and confer it on others. History records the price the Jewish people have paid for their unique heritage and lineage. I am just proposing that they were given this identity, and the offer to become the bride of Messiah, first. Their job was and still is to be a witness to the rest of the world. However, this authority to be the ambassadors of God has now been extended to all who believe, whether they are Jew or Gentile.

Thus God's original offer still stands. By believing in and relying on the work and person of the Messiah we can all be reunited back to our Maker. So, these two branches form two schools of fish, each with a rich, unique heritage. Yet both groups are measured and found wanting. Cetus is more than happy to carry us away, but the Lamb has paid the price for all of us. So — who are you swimming with today?

Andromeda Is Straining

The decan that follows The Band is *Andromeda*. Her story builds onto the ideas presented by the bound fish, for she has chains attached to her ankles and wrists. Andromeda's name, in Hebrew, means *The Chained*. Just as the two fish are banded and captured, so mankind is represented again as people in bondage who have no means of escape on their own.

We have also learned that sometimes the orientation these decans and constellations have with other decans in other groups can reveal additional insights. So it is with Andromeda. She is standing on the shoulders of Perseus, a decan still to be studied. Perseus is a warrior who has won a great battle. With a sword in his hand he has cut off the head of his foe. Meanwhile, the chains of Andromeda appear to be cut. Thus, with the help of Perseus she has been lifted up and has overcome her bondages.

Cepheus, the third decan in this group pictures a king sitting on a throne holding a scepter. To the right sits his queen Cassiopeia, another decan from another group. It is the perfect picture of the true King who will come. Even though we have been captured and put in chains, a King with the authority to cut our bands will come and rescue us.

The Constellation Aries and Its Decans

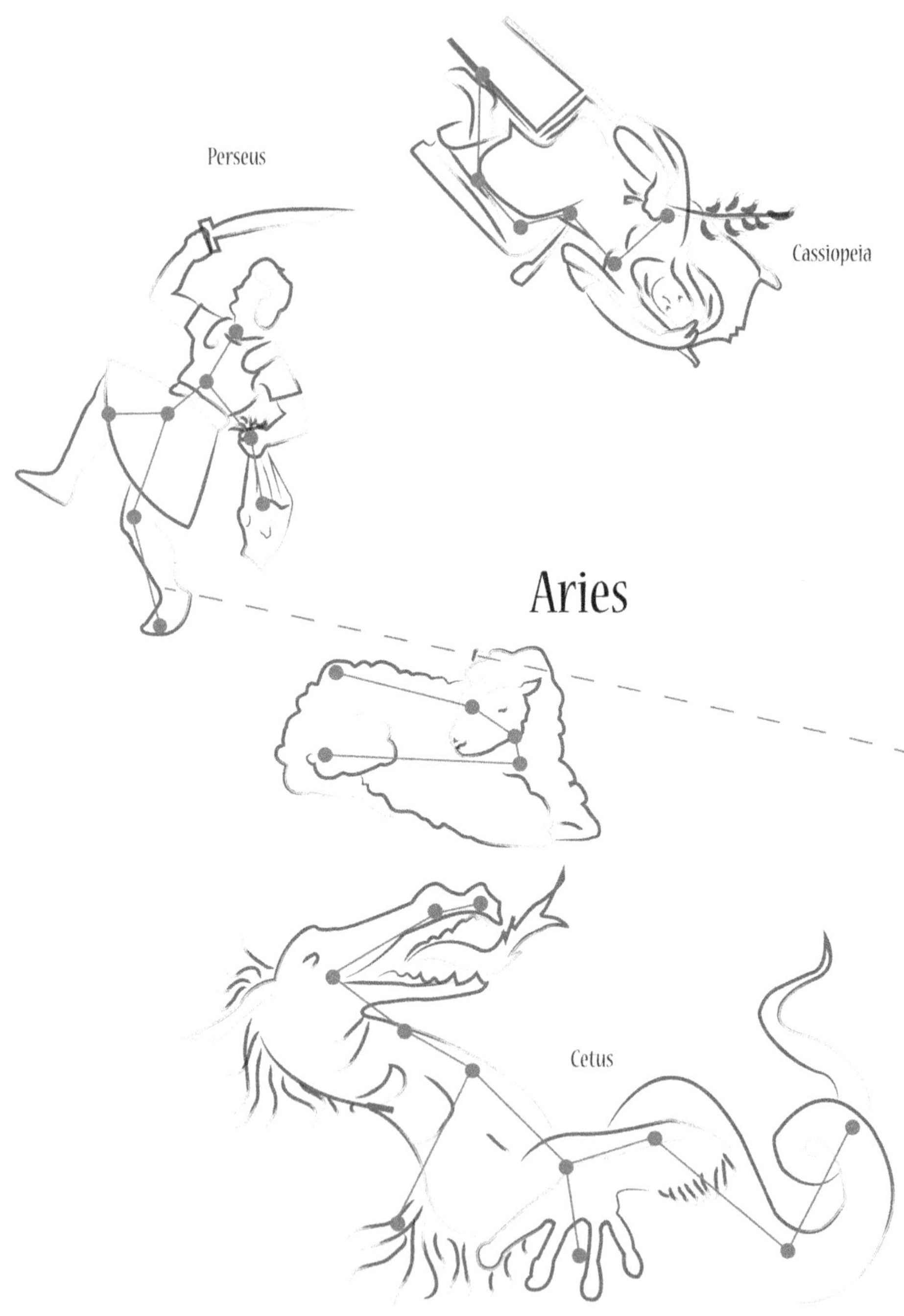

Act Two, Constellation Group Four: Aries

Aries is the eighth constellation group to appear in the procession in the sky. It takes its place overhead at midnight in the month of November. Libra, the scales, accompanies the rising of the sun during this same month.

This is the last of the four groups in Act Two. This Act has fully revealed the purpose of the Son's coming, and the reason for His battle and death. The first Act concerned itself mostly with the nature of the Son, and told us that He would come and die. Aries concludes the answer to the question we asked earlier. What was the motivation behind the Son's commitment to come and die on the cross? We discovered that His love for mankind drove him to make this sacrifice. And, that He has plans beyond that, too.

Meanwhile, this Act again revealed mankind's enemy once more, this time in the form of Cetus the sea monster, whom we mentioned before in the previous constellation group. In Act One, Satan appeared as a snake, a dragon, or a scorpion. All of these are animals that can be harmful, if not deadly, to man. The sea monster reveals Satan's attempt to be a bit more subtle, for he can hide his intentions better if he is beneath the waves of the sea, poking his head above the water only when he must.

We also met Aries the Lamb in the preceding constellation group. There he was intercepting Cetus as he was about to swim away with the two fish. The Hebrew name for this constellation is *Taleh*, which means "lamb." Recall how the Israelites were instructed to sacrifice and prepare a lamb for their Passover meal, just before they escaped from their captivity in Egypt 3,500 years ago. God asked them to celebrate this feast forever, as a remembrance of their newfound freedom. Little did they know that the lamb they killed and ate represented their Messiah, Who would someday come and offer His life as a substitute for the sins of all mankind so we could find our freedom as well.

This sacrificial lamb also represented the lamb that gave up its life to provide Adam and Eve their covering, thus relieving the pain of their disobedience. The biblical Messiah, Yeshua, was this very Lamb.

The *Taleh* is placed in the sky between two other decans, Cetus the sea monster and Cassiopeia, the bride who becomes Yeshua's queen. *Perseus*, the final decan in this group, stands as a warrior holding the head of our age-old enemy, Satan.

When God goes to battle in this final conflagration He will confront all the evil spirits, whether devils or demons. He will strike them down and cut off their heads. In Hebrew, the head represents authority. Thus God will once and for all remove all authority from these fallen spirits. No longer will they have power over mankind to deceive, maim, and destroy. Finally, at the end of the thousand-year reign, Satan will be released once more. But God will now deal with him for the last time, sending him to his eternal resting place, the Lake of Fire.

It is sad that so many will succumb to the wiles of this fallen cherub at the end of the thousand-year reign. Without Satan there to tempt them, only their own natures will be to blame this time.

Conclusion . . .

Meanwhile, as the first Act focused on the two enemies and the battle that would ensue, Act Two gave us the answer to the question, "What's all this fighting for?" Even as the two continued to battle it out, the prize they were fighting for was revealed.

Mankind, represented by fishes and Cassiopeia, is being saved from our enemy, the dragon who became a scorpion and has now become a sea monster. God chose to use these three images to teach us that the same natural abhorrence we have for these creatures should cause us to respond more aggressively and decisively to our first enemy, the one we met way back in the Garden.

The Constellation Taurus and Its Decans

Chapter Twelve

The Story in the Sky: Act Three

Act Three, Constellation Group One: Taurus

With the constellation Taurus we begin the final act of our story. Our search to discover the meanings of these voices in the heavens has been fruitful. Certainly, Psalm 19 comes alive and describes our experience so far. The heavens do tell of the glory of God. They do pour forth speech and reveal knowledge. And there is no place on earth where these voices cannot be heard, unless you live in a cave.

So, let's take a look at the last Act. This final section of constellation groups should bring us to the conclusion of our story. It should tell us more about mankind's destiny, where Satan ends up, and how his end comes about. If all this continues to line up with the biblical plan, then we can conclude that these stars and the Bible were originally created by God Himself, for no other person or group of writers could have been so accurate without His infinite knowledge.

This is one of the great truths that lead people to believe in the God of the Bible. No other religious book in the world contains prophetic material to the degree that the Bible does. And certainly, about 99 percent of the prophecies the other books do contain have been false. God really stuck His neck out when He gave His book to mankind.

On the other hand, the God of the Bible is no mere human. He knows the beginning from the end and everything in the middle, too! In fact, I believe He put so much prophecy in the Bible to function as a series of huge, glaring signposts for us when we read it. Perhaps God wanted to help us realize that there always was and always will be something very different and unique about its author . . . Himself.

The Ninth Constellation

Taurus is the ninth constellation group to appear in the procession in the sky. It takes its place overhead at midnight in the month of December. Scorpio is the scorpion accompanying the rising of the sun during this same month. Taurus is a picture of a bull charging head first. The back half of the bull morphs into Aries, the lamb we met in the previous Act.

Both of these figures represent Yeshua, our Messiah, who has promised to come twice: once as a lamb to offer Himself on the cross as payment for our sin, and the second time to come as a charging bull and become the permanent King over His creation. This will be accomplished when He defeats Satan for the final time and rewards those who have chosen to run the race to win. In I Corinthians 3:13–15 Paul spoke about our works and how they would be tested by fire. The outcome would determine our reward, but not our salvation.

As we will learn, the biblical image of a bull is used to represent the Messiah. Taurus, the charging bull, is coordinated with the decan *Eridanus* — the starry picture of a stream of fire, depicting the ultimate destiny of Satan at the hands of God, as foretold in the Bible.

One of the Old Testament sacrifices, the sin offering, suggests some parallels with the purpose of His first coming to earth. When the sin offering was made the bull was slaughtered in a very humane way, as all sacrifices were at the temple of God. The blood, the fat, the kidneys, and the liver were placed on the bronze altar and consumed by fire. These specific organs are all involved with cleaning and purifying blood.

The rest of the animal's body was taken outside the camp where the ashes from other sacrifices were disposed of. The bull's body was burnt up there.

Taurus is predicting how the real bull (Yeshua) would give Himself up to be our sacrifice. He was beaten and wounded — and bled extensively — inside the walls. But later He was led outside the city walls to offer His life on the cross.

The two separate sacrifices, one in the temple and one outside the camp, may very well be depicting the two-part nature of God Himself when He came as our Messiah. He was truly man but also truly God. The sacrifice done on the bronze altar depicts the acceptable purity of God offered by His Son, whereas the sacrifice outside the camp represented the sinful nature of mankind that He took onto Himself.

Par, the Hebrew word for bull, is spelled *pey, reysh*. Pictographically it means "to speak for mankind." Hebrews 5–8 informs us that Yeshua has now become our High Priest. As we know, one of the High Priest's functions was to stand between God and man and make petitions to God on behalf of mankind. To approach God in the Holy of Holies was a death sentence to all but the High Priest, and his approach occurred only once a year. But now Yeshua speaks on our behalf, serving as our High Priest in the order of the High Priest Melchizedek.

In the ancient world, the horns of a bull were sometimes placed on the coronet the king wore as a crown. It was a sign of power and authority. In fact, the bull was recognized as a symbol of a god. Many bull figurines, made of stone, have been found in the ruins of ancient peoples.

In the Mideastern world the bull idol also represented the god Baal. Baal was worshiped by the Phoenicians, better known in the Bible as the Canaanites. The Canaanites were the descendants of Canaan, who was the son of Ham. Recall that after Ham sinned against his father, Noah cursed Canaan, Ham's son. Canaan's descendants then migrated down into the land promised to the Hebrews. They were the seven tribes that God instructed Joshua to exterminate from the Promised Land.

Baal ultimately was an idyllic representation of Satan and becomes another copycat attempt by him to imitate the true creator God.

The First, the Strong, and the Mighty

The first letter of the Hebrew alphabet is *aleph*. In ancient times, when the Torah was written, this letter took the form of the head of a bull. It is not accidental that its pictographic meaning is "first, strong, and mighty."

The last of the Hebrew letters is *tav*. This letter was drawn in the form of a cross; pictographically it meant "the sign of the covenant." In the book of Revelation (and other places as well), God refers to Himself as the First and the Last, or the aleph and the tav. He truly is the strong and mighty sign that was hung on a cross. He made a covenant with us fulfilling an ancient biblical promise. The question is, will we respond back in covenant as well?

Certainly, God placed this bull in our constellations to remind us of His work on the cross and to give us hope. His ultimate goal will be to come back for His bride, mankind. This image too is copied by Satan. The stone bulls that represented Baal usually had the form of a naked woman on their backs. This female form was recognized as representing Astarte (from whom we get the word Easter), who was the ancient Babylonian fertility goddess. This figurine suggested a union between Satan and his wife, again copying God's plan to come back for His bride.

An Ox Is Not a Unicorn!

Unlike modern domestic bulls, Taurus almost certainly represents a now-extinct relative of domestic cattle, called *rimu* in the Hebrew Scriptures. Rimu is translated as "unicorn" in the Authorized Version and was once thought to be a mythological, one-horned creature. It is now known to be a larger and fiercer type of cattle, which modern translations usually call a "wild ox." Famous for its size and ferocity, it was the prize of great hunters in the records of Egyptian kings such as Tutmose III, and Assyrian kings as well.[20]

> Unfortunately, the unicorn mistranslation has been perpetuated in various versions of the Bible. In turn, what can seem like biblical support for the idea that unicorns once existed has been used to supposedly "prove" that the Bible is not accurate. In reality, unicorns are not real animals, extinct or extant, and the Bible does not actually claim anything to the contrary.

The First Decan, Orion

Orion is the first decan of the constellation Taurus. Here a great hunter is seen holding a club in one hand and the skin of a lion in the other. His name in Hebrew is *chesil*, the root of which means "strength or boldness." Three times this decan is named in the Bible, first in Job 9:9, 38:31, and then again in Amos 5:8. Certainly Orion conveys the image of strength and boldness as he holds up the lion skin in triumph.

The Second Decan, Eridanus

The next constellation, Eridanus, depicts another river. But in contrast to the river of life that flows from the urn held by Aquarius, this time it's a river of fire. The contrast is stark. This river is not a blessing at all, but a curse or a judgment.

Recall the oft-repeated biblical admonition to live a holy life.

> . . . but like the Holy One who called you, be holy yourselves also in all your behavior; because it is written, "YOU SHALL BE HOLY, FOR I AM HOLY."
>
> —I Peter 1:15, 16

> . . . so that you may remember to do all My commandments and be holy to your God. (Numbers 15:40)
> "You shall consecrate yourselves therefore and be holy, for I am the LORD your God."
>
> —Leviticus 20:7

> "Thus you are to be holy to Me, for I the LORD am holy; and I have set you apart from the peoples to be Mine."
>
> —Leviticus 20:26

The above is an interesting set of verses. Many more could be included, but these make the point. God's living water is for those who desire to be holy. His people's reborn identity comes from setting themselves apart from the ways of the world and acting and thinking like one who copies the ways of God.

Sandwiched between the last two "be holy for I am holy" verses from Leviticus is a list of commandments that serves as a definition of exactly how to be holy. The 1 Peter verse is quoting from these same verses in Torah. All are admonishing us that God's people are to be holy by ordering our lives according to a set of clear and distinct principles. These principles for holy living are reinforced again by Moses, one last time, just before he dies. In Deuteronomy 28, through Moses, God promises to give Israel blessings:

> "Now it shall be, if you diligently obey the LORD your God, being careful to do all His commandments which I command you today, the LORD your God will set you high above all the nations of the earth. All these blessings will come upon you and overtake you if you obey the LORD your God."
>
> —DEUTERONOMY 28:1, 2

Unfortunately for some, this verse is followed by another one that talks about what happens if people do not adhere to the ways of God.

> "But it shall come about, if you do not obey the LORD your God, to observe to do all His commandments and His statutes with which I charge you today, that all these curses will come upon you and overtake you."
>
> —DEUTERONOMY 28:15

It is not accidental, then, that when we get to the end of God's revelation about His plan, when all have had their chance to make their choices concerning the ways of God, the Bible ends with the book of Revelation. As we have learned from the *Lost in Translation* series,[21] this last book in the Bible brings to fulfillment the promises of Deuteronomy 28. Yes, Revelation reiterates these exact blessings and curses, spoken so long ago by God to His people. But so do the messages in the sky.

Embedded within the messages of the two rivers, one of water and one of fire, is the revelation and the fulfillment of God's promise to bring blessings or curses on mankind. All of us living today would be well served to heed these messages.

One Who Is Set Apart

Yeshua was referred to many times in the Bible as Jesus the Nazarene. His followers were also called the sect of the Nazarenes. The name *Nazarene* came from the town that He grew up in, Nazareth.

> "For we have found this man a real pest and a fellow who stirs up dissension among all the Jews throughout the world, and a ringleader of the sect of the Nazarenes."
>
> —Acts 24:5

Nazar is the root word for Nazarene, and Nazareth means "to be set apart or to separate oneself." It also carries the idea of consecrating oneself, taking a vow. The followers of Yeshua were really being called out as "set apart" ones, or the "people of the vow," fulfilling the idea of God's original intent. We set ourselves apart by holiness, by living out our lives according to biblical principles. It may sometimes be difficult, but living a principled life does indeed separate His people from the crowd. God has promised a great blessing for those who do. We don't separate ourselves from the world physically, but by our actions and lives the world will be able to tell the difference.

These two streams fulfill the promises of the blessings and the curses given to mankind approximately 3,500 years ago.

> A fiery stream issued and came forth from before him: thousand thousands ministered unto him, and ten thousand times ten thousand stood before him: the judgment was set, and the books were opened.
>
> —Daniel 7:10, KJV

Eridanus, as described by Daniel above, exemplifies the destiny of the ones who have chosen to go their own way. Satan chose this path ages ago. We follow him at our own expense.

The Third Decan, Auriga

Auriga, the third decan of Taurus, represents those who have chosen the blessings. This constellation pictures a shepherd holding a she-goat and its two young lambs on his lap. It is a perfect picture of the One who has promised to be a protector and lover of His people. It completes the idea of consummation, of bringing to a close God's promised plan.

The victor comes in strength and deals a mighty blow to the devouring lion as depicted in the decan Orion. Then He separates His chosen — those who have identified themselves by walking in God's way from those who have not. The organization of these pictures in the sky reveals how the bull and the mighty hunter collide, with the lion skin between them.

This scene separates the two other decans in this group. Above lies the blessed on the lap of the Shepherd. Below flows the river of fire leading the cursed ones to their distinctly different destiny. At the end Satan will be separated from the blessed, no longer having access to their hearts and minds, and he will be a trouble to them no more.

God in His great wisdom has used the sign of a bull to represent Himself. In doing so He exposed another of His attributes. He is a leader with strength, One who has nothing to fear. Even the lion fears attacking a fully grown bull.

The Hebrew aleph-bet is phonic like the English alphabet, but it is also pictorial. Each of the Hebrew letters represents a concept communicated by a picture. The first letter of the Hebrew aleph-bet is, of course, an aleph. Today its shape does not clearly suggest its original idea, but in ancient times it resembled the head of a bull or an ox.

Even though its picture has evolved away from its original form, it still retains its original meaning. In ancient times a bull or ox conveyed the idea of a mighty one, the one who is first. This makes additional sense because the aleph is the first letter. But being also a picture of a bull and combining the ancient symbology of a bull representing Elohim, the biblical title for God, the ancient aleph is conveying something important about the Creator. When these biblical and pictographic concepts are combined with the pictures in the sky, Taurus the bull can be clearly identified for who and what he really represents. He is the God of the Bible who will come as prophesied, to put away the enemy (the snake, or Satan) and protect and draw to Himself His prize (mankind).

The Constellation Gemini and Its Decans

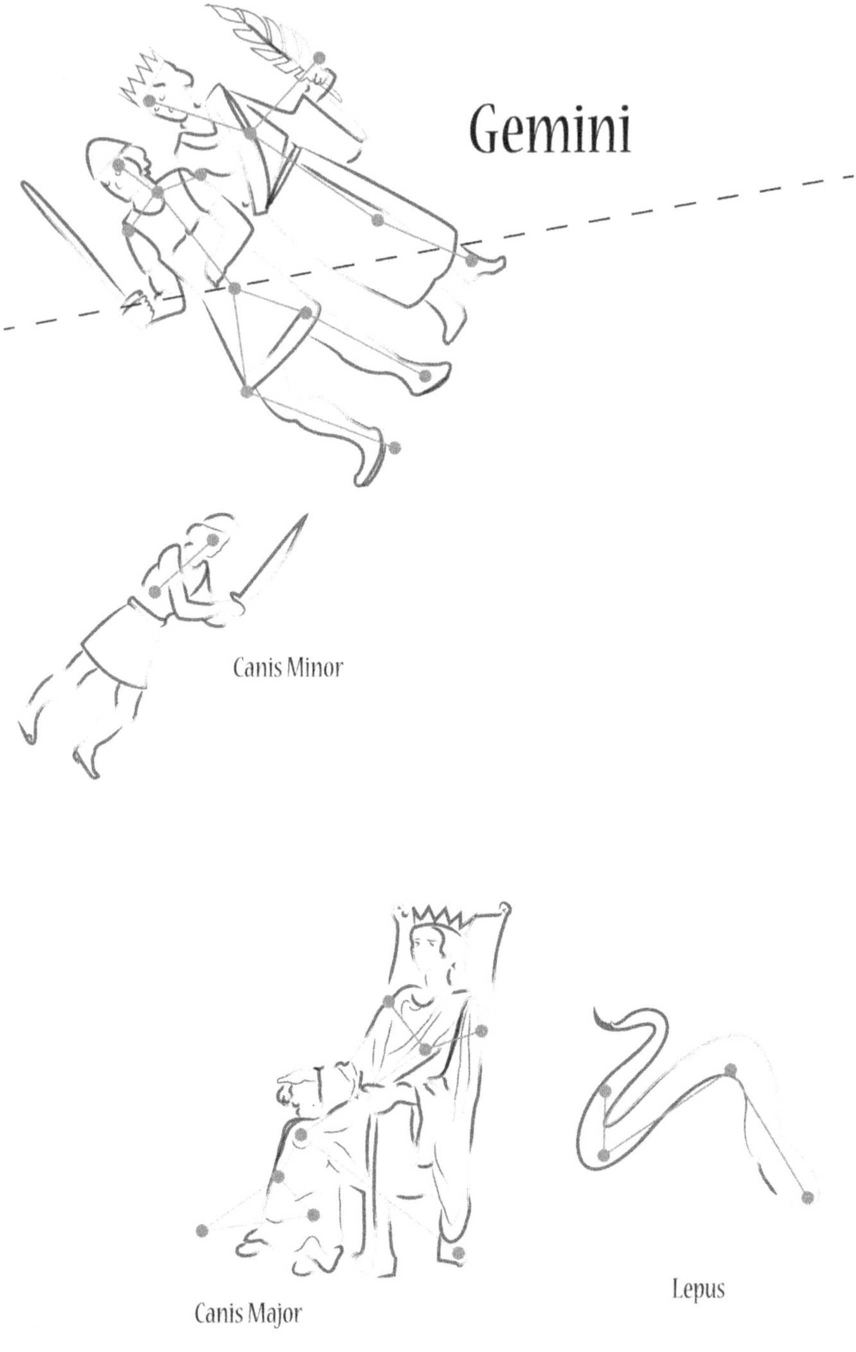
Gemini
Canis Minor
Canis Major
Lepus

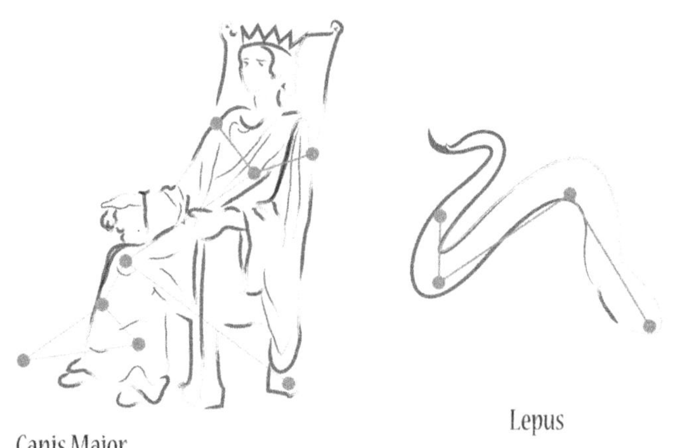

Act Three, Constellation Group Two: Gemini

Gemini is the tenth constellation group to appear in the procession in the sky. It takes its place overhead at midnight in the month of January. Sagittarius, the centaur, accompanies the rising of the sun during this same month.

Gemini means "twins" and is usually a picture of two babies or two men. More rarely, the two are depicted as a man and a woman. The ancient pictures of these two show one holding a palm branch with a crown on his head while the other holds a sword.

Pollux and *Castor*, the two brightest stars in this constellation, help us understand the underlying biblical message. Pollux means "Who Comes to Labor or Suffer" whereas Castor means "Ruler" or "Judge." These stars make it easy for the Bible believer to see that these twins are revealing the two visitations of God as prophesied in the Bible.

The first coming, foretold in many places in Scripture, is depicted by Pollux and is represented by the twin holding the sword. When Yeshua came the first time He told His followers that He did not come to bring peace but would cause division. With his sword this twin did exactly that. The world has been fighting over God and His Word ever since His first coming two thousand years ago. When the Tanakh (Old Testament) describes God's coming for the first time, He is referred to as the suffering servant, paralleling the meaning of Pollux, the star whose name means the same.

Meanwhile, anyone familiar with the prophecies pertaining to God's second coming knows that instead of coming to suffer again on the cross He will next come as a conquering hero and king. His primary image thus becomes a ruler and judge, which is the exact meaning of Castor, the second of the bright stars in Gemini. The second twin is holding a palm branch and is wearing a crown. Thus the idea of a ruler is suggested again, as well as that of a judge who will bring peace upon the earth. Peace will come because Satan and his followers will be judged and removed, and those who remain will be instructed in the ways of holiness by God.

The Hebrew name for this constellation is *Thaumim*, which is a derivative of *thummim*. Recall that the Urim and Thummim were instruments to be put in the breastplate the High Priest wore. This would position these items over his heart. Their purpose was to bring insight into the ways of God. But the underlying Hebrew meaning of these words reveals more about the nature of God and the way He works. "Integrity," "wholeness," and "truth" are the meanings of thummim, while the word urim means "lights" or "revelations."

Thus the High Priest was to serve God's people even as he observed these foundational principles. God's instruction through these ancient ideas is the same today. It is still wholly applicable and, in fact, is needed now more than ever. As we walk in integrity and in truth God will light our way and give us instructions and reveal His path to us. Those who meditate on these principles will be blessed.

> But his delight is in the law of the Lord, and in His law he meditates day and night [3] He will be like a tree firmly planted by streams of water, which yields its fruit in its season and its leaf does not wither; and in whatever he does, he prospers.
>
> —Psalm 1:2, 3

The word in the above passage, for "law," is the Hebrew word *Torah*. Throughout the Bible God is represented as truth and light.

> Jesus said to him, "I am the way, and the truth, and the life; no one comes to the Father but through Me."
>
> —John 14:6

> This is the message we have heard from Him and announce to you, that God is Light, and in Him there is no darkness at all.
>
> —1 John 1:5

> In the beginning was the Word, and the Word was with God, and the Word was God.
>
> —John 1:1

In the first chapter of Psalms, quoted above, by using the word "law" David is referring to the primary meaning of the Hebrew word Torah, the first five books of the Old Testament. These contain the principles and instructions we should be meditating on, for these truths are the very ones that are not abolished but will be fulfilled.

> "Do not think that I came to abolish the Law or the Prophets; I did not come to abolish but to fulfill."
>
> —MATTHEW 5:17

The above verse in Matthew then tells us that not even the smallest letter or the least stroke of a pen will be eliminated from the Law until everything is accomplished and heaven and earth disappear.

This constellation's Hebrew name is communicating to us the principles on which God's work and His plans for the future rest. We can be assured that no untoward or evil motive will ever lie behind the purposes of God. As this constellation suggests, He will come to earth twice. Each time, part of His plan will be accomplished. Integrity and truth will be the cornerstones of His purposes and His reign.

The constellations of Act Two also tell us why the Son of Virgo would come and give up His life. The reason is to save you and me! We can be both assured and comforted by the messages of the stars and the Bible, because they agree and are consistently true wherever they are found. It is not accidental that the Bible and the messages of the stars align.

The First Decan, Lepus

Although *Lepus,* our first decan in Gemini, is pictured as a hare in modern times, originally he was a snake. Lepus' brightest star, *Arneb*, means "The Enemy of Him Who Comes." That the enemy of Gemini (Yeshua) is Satan, commonly depicted in our constellations as well as in the Bible as a snake, only confirms that Lepus should not be a picture of a member of the rabbit family.

Considering that Lepus is located way below the feet of the twins, and just below the feet of Orion, the decan of Taurus the bull, the prophecy in Genesis 3 comes to mind. The twins and the mighty hunter, Orion, all reflect attributes and authorities of God Himself. Recall that this Genesis passage communicates to us that a Son of Eve will destroy the snake by wounding it on its head. Knowing that in Hebrew the head represents one's authority, the location of this wound foretells the destruction of Satan's temporary kingdom. We know that kingdom will end at the coming of the second twin, Yeshua Himself, who will conquer and reestablish His right to rule and judge His creation.

The next two decans in this group are *Canis Major* and *Canis Minor*. They are recognized as Orion's two dogs; thus "canine" is the genus name for dog, wolf, and coyote. One of these dogs was greater and one lesser. However, many believe that both of these pictures have been changed through the centuries. The inconsistency of their forms in each culture, and the star names themselves, again suggest that their original forms were not dogs but rather each one of the twins in Gemini, separately.

The star *Sirius*, the brightest star in the heavens, is found in Canis Major. Its name means "The Prince." The ancient Hebrew root for this word is *sar,* which means "prince." It survives in its English form today when we say, "Yes, sir!" This Prince represents the twin in Gemini, who comes to rule and usher in everlasting peace.

The Egyptians called Canis Minor *Sebak,* which means "The Conquering/Victorious." This reveals that this decan, instead of being pictured as a dog, should take the form of the other twin of Gemini, the twin who comes to earth to do battle with the snake but also to die on our behalf.

Through integrity and truth God will visit the earth and confront Satan. In the end God will take up His rightful place, displacing the snake from God's temple. The story this constellation and its decans tell is about bringing things to a close. The last two groups complement the idea of bringing to a finale the plans of God. Let's see if it ends with an exciting climax. I'll bet it does!

The Constellation Cancer and Its Decans

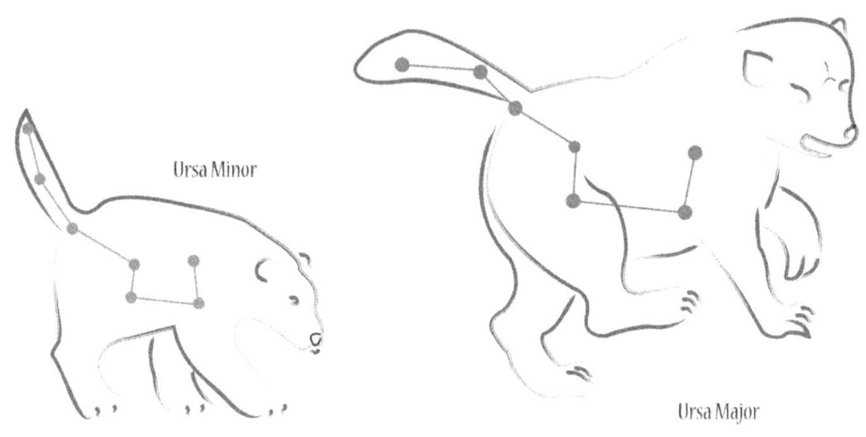

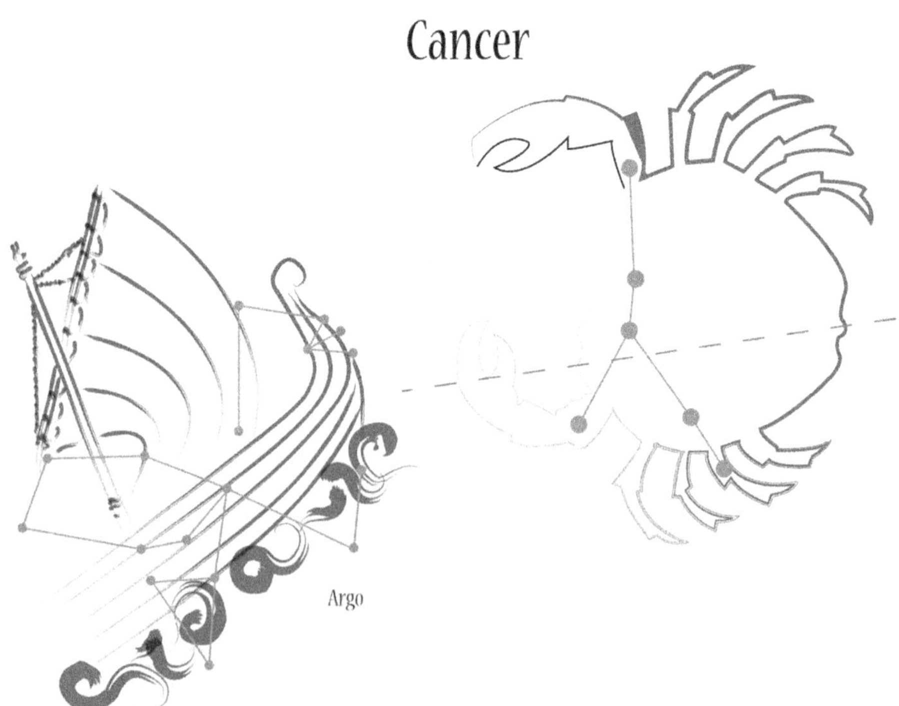

Act Three, Constellation Group Three: Cancer

Cancer is the second-to-last constellation and is the eleventh constellation group to appear in the procession in the sky. It takes its place overhead at midnight in the month of February. Capricorn, the goat/fish, accompanies the rising of the sun during this same month.

Cancer takes the form of a gigantic crab, which lies between Gemini and the lion of Leo. It is seen as a crab in the Parsi, Hindu, and Chinese zodiacs. However, in the ancient Dendera zodiac, in Egypt, we see the scarabaeus represented, which is a sacred beetle. The sign is the same in the zodiac of Esneh, as well as in a Hindu zodiac dated 400 BC. Its name is *Karia*, meaning "the folds or the resting places." The Arabic name is *Al Sartan*, which means "he who holds or binds."

In Hebrew there is no word for the crab, which was classified as vermin and an unclean animal. The Greek word is *Karkinos* while the Latin is *Cancer*, which both mean "holding or encircling." *Khan* means "the travelers resting place," and *ker* or *cer* means "embracing, encircled, or held as within encircling arms." The ancient Akkadian name of the month is Sukulna, meaning "the seizer" or "possessor of seed."[22]

Some speculate that the crab or scarab pictures have replaced the original constellation. They propose that the original might have been an ancient sheep pen made from rocks, piled up to form a circular fence. This pen would hold the shepherd's sheep securely and safely through the night when wolves might come to harm them. Whether these propositions are correct or not, even from the idea of the existing constellation pictures of a crab or scarab it is clear that a consistent meaning is revealed by all.

The Shepherd Who Breaks Down the Walls

In the King James translation of Matthew 11:11–12 we find the following:

> Truly I say to you, among those born of women there has not arisen anyone greater than John the Baptist! Yet the one who is least in the kingdom of heaven is greater

> than he. [12] From the days of John the Baptist until now the kingdom of heaven suffers violence, and violent men take it by force.

Various commentators have misinterpreted the last sentence, above, for hundreds of years. The Amplified Bible even adds the words "as a precious prize — a share in the heavenly kingdom is sought with most ardent zeal and intense exertion" to the end of it. But in reality, these verses refer directly to a passage found in Micah 2:12–13:

> I will surely assemble, O Jacob, all of thee; I will surely gather the remnant of Israel; I will put them together as the sheep of Bozrah, as the flock in the midst of their fold: they shall make great noise by reason of the multitude of men. The breaker is come up before them: they have broken up, and have passed through the gate, and are gone out by it: and their king shall pass before them, and the LORD on the head of them. (KJV)

In the original Hebrew, the word translated as "breaker" is *peretz*, meaning either a shepherd who breaks boundaries to release his sheep ("breaching forth") or a woman going into labor (a birthing term).

Typically, an ancient Hebrew shepherd penned his sheep up each night, within an enclosure made of rocks. When it came time to let them loose in the morning, the shepherd (i.e., the "breaker") would knock out some of the stones in the wall with his staff. The sheep themselves would then expand the breach in the process of escaping, as they were set free or "birthed."

In a similar way, this passage explains that, in response to Yeshua, the kingdom of heaven is "bursting forth." It is clearly a Messianic reference; Yeshua is saying, "I smashed the rock out and birthed you forth; I am the shepherd who breaks down the walls and sets you free."

When the Greek words that are commonly translated as "the kingdom of heaven suffers violence, and violent men take it by force" are translated back into what was undoubtedly the original Hebrew, they become an idiomatic expression (as described above) that lines up perfectly with what Micah was saying.

In other words, the whole thing makes no sense at all in the standard Hebrew-to-Greek-to-English sequence. Only when the Greek is translated directly back into Hebrew, and then directly into English, does it ring true.[23]

> According to ancient Egyptian myths, the sun (Ra) rolls across the sky each day and transforms bodies and souls. Modeled upon the Scarabaeidae family *dung beetle*, which rolls dung into a ball for the purposes of eating and laying eggs that are later transformed into larva, the scarab was seen as an earthly symbol of this heavenly cycle. This came to be iconographic, and ideological symbols were incorporated into ancient Egyptian society.[24]

The above quote from Wikipedia adds to the ideas of surrounding and protecting. The scarab is a picture of God. God provided a means of sustenance and transformation for our bodies and our souls via both the sun, above, and His own Son, Yeshua. The crab is prophesying that God's plans conclude with us as the apple of His eye. Because of His love for us He has promised to protect us and bring us to Himself.

Thus we can rest in the knowledge that God has everything under control when it comes to us, and everything else too, for that matter. Gemini reveals who the King is, and who is the winner of the galactic, age-old battle between God and Satan. Cancer reveals the reason for the battle in the first place, and the part mankind is to have in God's plans. His end game is to protect us with His surrounding love.

Certainly this idea of surrounding in an effort to protect is seen many times in the Bible, when God reveals His love and purposes for His people. God is described as having wings of an eagle, coming to deliver love and care, then taking His chosen to a place where he protects them from their enemies. Deuteronomy 32, Exodus 19, and Matthew 23:37 all say the same thing, as does the following quote from Revelation.

> And when the dragon saw that he was thrown down to the earth, he persecuted the woman who gave birth to the male *child*. [14] But the two wings of the great eagle were given to the woman, so that she could fly into the wilderness to her place, where she was nourished for a time and times and half a time, from the presence of the serpent.
>
> —Revelation 12:13, 14

Wings of eagles, the arms of a crab, or the animal pen of a shepherd — all reflect the heart of our Creator and this constellation's original meaning, whatever the ancient picture actually was.

The First Two Decans, Ursa Minor and Ursa Major

The first two decans are *Ursa Minor* and *Ursa Major*. Of course we know them best as "Little Dipper" and "Big Dipper." Lesser known names today are "Little Bear" and "Great Bear." However, the forgotten meanings of these names are the "Lesser and Greater Sheepfolds."

Star names such as "The Guarded, The Numbered, The Visited, The Multitude of the Assembly, The Ransomed, The Redeemed, The Purchased," and "The Flock" all bear witness to the very heart of God toward His people. His goal is to protect them just as a shepherd would assemble and corral his sheep in a fold for the night against wolves bent on destruction.

Ursa Minor's brightest star is Polaris, also known as the "North Star" or the "Guiding Star." It takes its place in this decan at the very point of its handle. However, Polaris, our pole star today, has not always occupied that central place in the sky. Six thousand years ago, the approximate time of the fall of man at the hand of Satan, the constellation Draco, "The Dragon," held this place of honor. At that time, all the host of heaven and all the constellations revolved around Draco, who best represents our enemy, Satan.

It seems that God knew the heart of man and the choices he would eventually make even as He was creating our universe. He knew we would choose to have our lives revolve around someone other than Himself, someone who would offer prizes that would feed our lusts, our pride, and our inclination to make ourselves the most important. The heavens revealed that Satan would make himself ruler over the creation. He would want everything to revolve around himself.

However, we have some bad news for Satan and some very good news for those not caught up in Satan's vortex. Over the centuries the pole has migrated from Draco to Polaris. God built into the earth's night-time prophecy that even though Satan would rule the creation for a while, in the end God would take back what is rightfully His. As Polaris takes its place of authority in the handle of the Little Dipper, as a hand would grab the handle on a cup, so God will come back to claim what is rightfully His.

The two stars that form one of the sides of the Big Dipper also involve Polaris by precisely pointing to Polaris' position in the sky. The star names that make up these dippers suggest that they are the sheep in the fold being protected by the Shepherd.

One may ask why God chose to organize His people into two groups as reflected in the two dippers, instead of only one. In Romans 1:16, as well as several other places in the Bible (e.g., Romans 2:9, 10), God identified two groups of people when talking about the recipients of salvation.

> For I am not ashamed of the gospel, for it is the power of God for salvation to everyone who believes, to the Jew first and also to the Greek.
>
> —Romans 1:16

Could these two cups (the two dippers) be representations of these two groups? The lesser cup could represent the Hebrews, who were called *Yeshurun*, a tender and loving appellation, which was used four times in reference to Israel (Deuteronomy 32:15, 33:5, 33:26, Isaiah 44:2).[25]

Have Enough Integrity

The Hebrew root meaning of *Yeshurun* is "to be upright, to have integrity." It seems that even though God knew that His people would rebel and chase after their own lust by worshiping other gods, He still called them by this endearment. The Hebrew word for God, *El*, makes up the word *Israel* and means "the prince of God." When talking about all who are grafted into God's family through their individual faith, whether Jew or Gentile, He consistently refers to them as *Israel*.

This is how He sees His people. Even though we sometimes fall He wants us to have the integrity to get ourselves up off the ground, ask for forgiveness, turn from our old ways, and fight the good fight.

Maybe the Big Dipper, because it includes a larger group of people, represents the Gentiles. God positioned Himself, Polaris, as a star forming the decan Ursa Minor. This may suggest His bloodline, which we all know was Jewish when He came in the flesh. It may also support the idea that, in God's mind, the Jews were first as the above verses teach.

However, I do not mean to suggest that God was implying *value or importance of the Jewish people themselves* by using the word "first" in these passages. Rather, I believe it reveals the fact that He gave the Jews His principles first. It also reveals the order of responsibility to be a witness to the world as they worked out these truths in their own lives (Deuteronomy 4:1–8).

Whatever these associations mean, something interesting is revealed by the star Polaris. Today, as stated above, Polaris is the pole star, meaning that all the stars in the night sky revolve around it. However, as its lesser known name ("Guiding Light") suggests, mankind has used this star to direct his way home. Anyone familiar with the skills of navigation knows that this star is fundamental in finding one's way at night. Certainly, Yahweh wants us to plan our ways, our decisions, and the principles we use to guide our paths around Him. His light should guide us in our thoughts and actions as we navigate each day.

The question, then, is why are so few choosing to do so? Most seem to be walking away and not paying attention to the instruction of the voices from above.

But there is something more. Polaris is not a single star but rather a star system. It includes a bright, super-giant star which is the one we can see clearly with our naked eyes. But orbiting very closely are two additional stars. These two are much smaller in size, and are virtually invisible to anyone not using a telescope. Even then, they are seen traveling around this super giant at a very close range. What does all of this mean?

These stars may be suggesting that there is only one God, as His Word suggests (Isaiah 45:5, 6, 21; Mark 12:32; James 2:19). Likewise, we can only observe one light with our unassisted vision. However, close inspection of Polaris reveals God's nature. As Genesis 1:26, 27 suggests, God refers to Himself by using plural pronouns in verse 26, saying, "Let Us make man in Our image." But in the very next verse He harmonizes His singular state of existence by saying, "So God created man in His own image," this time using the singular pronoun "His."

Those who concentrate on the New Testament are especially aware of the phenomenon of God's referring to Himself in three different ways. These three stars represent the Father, the Son, and the Holy Spirit. But man is also a composite being. We certainly have a body, but we also have a spirit and a mind/soul. That, however, should come as no surprise, for again, we are made in His image (Genesis 1:27). The triune nature of God is revealed in creation in the Pole star as well as in the construction of man, all of which confirms the biblical text.

Since God is about to return for His bride, she should be readying herself by orienting her life around Him. The movement of the stars and constellations, now around Polaris (God) instead of Draco (the Dragon), is a prophecy forewarning mankind that time is about up. There is a new King in town, and He will call on us to see what we have done with His Son and the book He has given us.

Different Titles, Different Times

Again and again God uses different titles for Himself throughout the Bible. Each time He reveals the voice of origin from which He wishes to speak. We do the same thing today. When speaking to our children we use a different voice than when we speak to our boss or our spouse. In life we have different capacities and authorities. Mixing them up can get us into a lot of trouble.

When speaking to mankind God uses different voices as well. Each voice will convey a different message. Personally, this is one of the reasons I do not like the translations today and I am driven to use the original Hebrew text. The individual names of God are many times not translated. From the English texts you cannot tell what voice God is using, whether it's His personal name, *Yahweh*, or a different title for Himself such as *Elohim*, which means "God."

The Third Decan, Argo

The last of the decans within the constellation Cancer is *Argo*. This name means "A Company of Travelers." Star names in Argo mean "The Possession of Him Who Comes," "Possession," and "The Desired." These names confirm the theme of the entire group of the eleventh constellation. God is coming back for His people, people who have been guarded and loved by Him. These travelers have been taken to a place of protection. The sheep pen and the ship riding through the rough waters remind us of His promise to come to us providing protection, on the wings of eagles.

In ancient times, each ship had a mast that faced forward in the direction the ship was heading. Each mast then had a carving of some animal or god positioned just below it. However, instead of having the head facing forward, Argo has the head looking back into the ship's contents, the travelers. The imagery can't be mistaken — no need to look ahead. God knows full well where He is going. Those reflected in His eyes are the ones who love Him and have accepted Him for who He is: their Groom.

At this point in the progression of the story in the sky we see our Groom coming for His bride. She comes riding through rough waters, but the next and last constellation tells of a short stretch of time that is uniquely reserved for Satan.

This order in the story confirms our findings with respect to the teachings of the book of Revelation. In the *Lost in Translation* series, my co-author and I explained that God comes for His bride before the last seven bowl judgments are poured out upon the earth. We learned that these tribulations will then last three-and-one-half years.

Will the bride be taken up to be with Him before all of the terrible times come? No. Some will suffer and be tried, even unto death before being taken to be with their Groom. Others will survive through these end-time storms. These sheep pens, encircling crab claws, and Argo, the ship carrying His bride to safety, clearly tell of a loving God, testing the love of His bride but delivering her through all the trouble Satan can throw at her.

The Constellation Leo and Its Decans

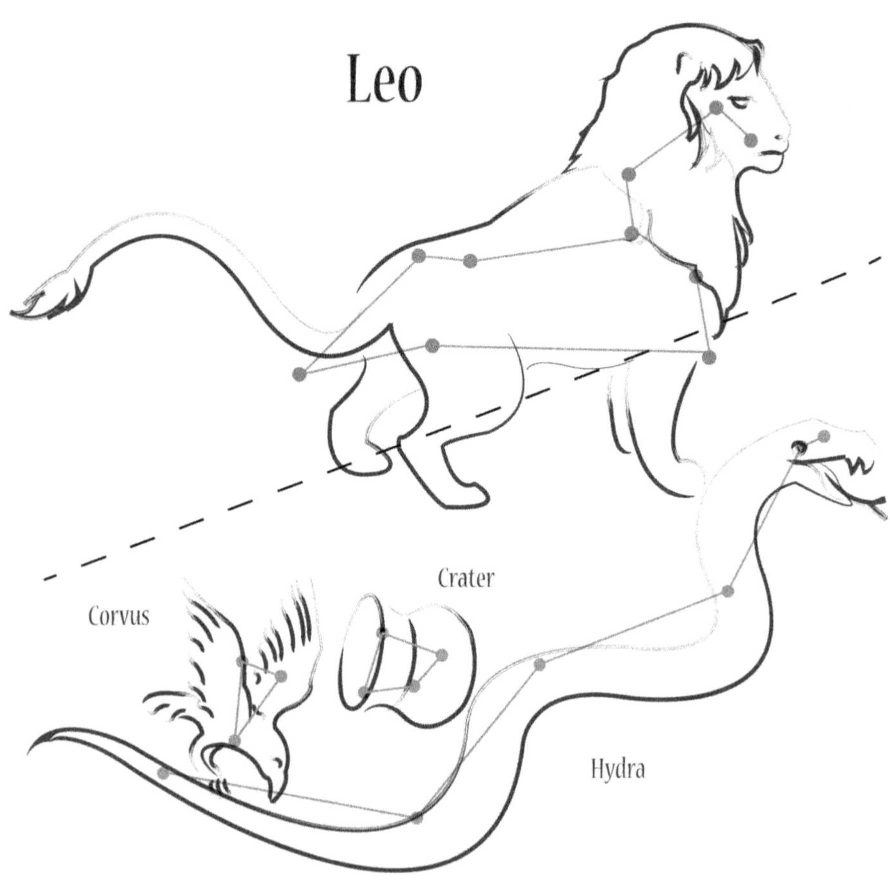

Act Three, Constellation Group Four: Leo

Leo is the twelfth constellation group to appear in the procession in the sky. It takes its place overhead at midnight in the month of March. Aquarius the water bearer, who is pouring out the life-giving waters, accompanies the rising of the sun during this same month.

This group provides us with the final message from the starry hosts, and it is one of the most profound of them all. This constellation group brings us to the climax of that age-old battle described to us by the story in the sky that we've already examined, between Elohim and His wayward but most beautiful creation, Satan.

This scene depicts God as Leo taking command, and Hydra, the serpentine enemy, receiving its reward for what it has "accomplished." There is no battle scene here. No dispute about who is the winner. No question as to who is in control of destiny. Only the snake and his reward, if you want to call it that, are left to be revealed.

> "No doubt whatever exists about the meaning of Leo. Long before the Greeks invented their myths about the exploits of Hercules, Leo was a well-known astronomical sign. Without exception, the oldest zodiacs picture Leo as a lion. Not only is he a lion, but a magnificent lion killing and tearing his prey. In the famous zodiac of Dendereh, Leo is treading down the great serpent, Hydra, his powerful claws having wounded the head of the serpent."[26]

The names of the stars within Leo confirm this identity: "Treading Underfoot," "The Judge Who Comes," "The Exaltation," "The Tearing of the Lion," and "Enemy Put Down."

Leo, well known as the King that He is, stands with His front feet on top of the head of Hydra. This is the fourth time we have seen this symbology in the constellations. Leo, Ophiuchus, Taurus, and Hercules — all pictures of mighty beasts or mighty warriors — have their feet upon the head of Satan, who is pictured as a snake or a dragon.

Hydra represents our ancient enemy, Satan, one last time. His authority, represented by his head, is being suppressed, destroyed, and then removed.

> "YOU HAVE PUT ALL THINGS IN SUBJECTION UNDER HIS FEET." For in subjecting all things to him, He left nothing that is not subject to him. But now we see not yet all things put under him.
>
> —HEBREWS 2:8

Paul, in quoting Psalm 8 above, confirms the image we see of Leo and the previous three constellations. Satan is not subject to the authority of our creator and savior Yeshua — at least not yet. In His great wisdom God has given Satan some latitude, all for the purpose of testing His people's love for Himself. But someday that righteous, godly authority will deliver this rebellious cherub to the place reserved for him: the Lake of Fire, as in Revelation 20.

The First Decan Is Hydra

Hydra is a huge decan. With its head under Leo's feet, its enormous length is exposed and stretched out around several additional constellations. Hydra encompasses about one third of the perimeter of the circle that the constellations form. Thus its length confirms this passage in Revelation 12:

> Then another sign appeared in heaven: and behold, a great red dragon having seven heads and ten horns, and on his heads were seven diadems. And his tail swept away a third of the stars of heaven and threw them to the earth. And the dragon stood before the woman who was about to give birth, so that when she gave birth he might devour her child.
>
> —REVELATION 12:3, 4

Remember that, in Hebrew, the word for star can metaphorically mean angels. Satan's tail does indeed sweep up one-third of them and throws them to the earth as it stretches around one-third of the sky. These are the angels who fell for the deception that Satan offered them, which was very similar to the one he offers us. *We don't need*

God; we can become lords ourselves and we can rule the earth and live our lives together without Him!

It is not by accident that so many people pay no heed to their real Lord. They make decisions without asking direction. They think and do things without giving any consideration to the guidelines He gave us. They do not recognize the blessings He has bestowed on them, instead thinking that everything results from their own efforts.

Satan has been the sorcerer behind the curtain all this time, but his time will end as prophesied by this ancient decan. His end is foretold by the head wound he is receiving (Genesis 3:15), but the final two decans also pile on, adding additional detail about what God has in store for Satan.

The Final Two Decans, Corvus and Crater

The final two decans are *Corvus* and *Crater*. Respectively, they are pictures of a raven and a bowl. It is fitting that God chose to bring to conclusion His panoramic story of restoration and redemption by using imagery right out of the last book in His Word, the book of Revelation. This last book describes the judgments God has in store for Satan, just before He comes back to earth to claim what is rightfully His. Those judgments are exactly the same as what Corvus and Crater promise. Crater, the bowl of judgment, is attached on Hydra's back. Meanwhile, some of the stars that are included in Hydra also form Crater.

It seems that, from the very beginning, when these constellations were conceived in the mind of God, His response to rebellion against Him carried with it curses and judgments. In fact, rebellion and death may be one and the same. To rebel from God is to separate yourself from Him. To separate yourself from God is to die. That is why God said to Adam and Eve that on the day you eat (rebel) of the Tree of the Knowledge of Good and Evil, you shall surely die.

In Revelation, these judgments are symbolized by seven bowls (Revelation 16), which are being poured out upon the back of the kingdom of the beast (Hydra). The symptoms caused by these bowls are very similar to the symptoms of syphilis.[27] The last of the seven resulted in death to the beast, and that is exactly the same result here in the stories of the stars.

Corvus the raven is filling its belly by picking out chunks of meat from the back of Hydra. Again, in Revelation 19, as the real King comes to take His kingdom back and defeats the beast and his army, birds are told to come and feed off the dead bodies of Satan's army. God is indicating to us that when He comes back a cleansing process will begin. As with all of mankind, the earth and all that has been made impure and unclean will be restored or removed.

These punishments fulfill the promises of God that He gave the world in the very beginning. Deuteronomy 27–28 tells of God's blessings to those who will love and obey Him, but it also details the curses that will result if His creation doesn't listen and goes its own way. Thus the book of Revelation brings to consummation God's promises. He has promised to bless those who choose Him and to judge those who choose otherwise. The book of Revelation presents two stories, two destinies. The voices of the sky proclaim these same promises — one of blessing, life, and salvation, and another of punishment and destruction.

The Final Act

This final act in our series of three reveals exactly what we would suppose. After proclaiming in the first two acts that a battle has been going on between two beings since the beginning of time, and that the two opponents are fighting over mankind, the important questions then come into focus. Who will provide a government for them to live under, who will be their king, and who will rule and reign over the creation?

These last star formations reveal the winner, as if it was ever in question. They also tell of the destinies of God's people, and of those who are not. And, of course, they also reveal the destiny of Satan.

The latter two receive God's punishment for not following Him, even as God's people receive the promised blessings — freedom, forgiveness, and the authority to rule and reign with Him. Thus these stars do, indeed, foretell a future destiny we can all hope for. And, that destiny aligns perfectly with the prophecies given to us in His Word.

None of this is by accident; it's by design. The ancients, and certainly Satan, could hear these voices in the heavens. Via Satan's intervention they have sometimes been hidden from our eyes and perverted, even as their messages have become scrambled.

But our study has uncovered the ancient meanings. The stars and the messages of the Bible do, indeed, align. They complement and build on each other, giving us hope and confirmation that both have been created by someone bigger and more powerful than ourselves.

To put one's hope in Him is not folly. It does not mean that we are blind or foolish. On the contrary, the more we learn about science, history, and the world around us, the more clear and distinct God's own voice becomes. He wants you and me, for we are His endgame and the reason for the battle in the first place. The way has been made for us; all we must do is step out earnestly and resolutely on the path provided.

Chapter Thirteen

The Feast Overlay

In chapter four we discovered that the seven feasts God instructed His people to celebrate each year actually foretold and revealed the plans God had for our protection and redemption. This plan has now been complemented, embellished, and confirmed by the messages we have found coming from the constellations. In their order of appearance do they also correlate with the timing of the feasts?

In this chapter I want to explore the timing of the feasts with the constellations that appear during their celebrations. Each feast was given a specific date on which the celebration was to occur. However, these dates were based on the ancient lunar calendar. God's lunar calendar and our western, solar-based calendar differ on the starting day for each month, but not by enough to cause a problem. So, for ease of reference I will use the months of the Gregorian calendar.

Let's start with the very first feast, Passover. This occurs on the 14th day of Nisan on the Hebrew calendar (the first of Nisan is the religious New Year). This date usually occurs somewhere between the last days of March and the end of April. I will use the constellation that is overhead during the month of April as our reference for Passover.

The constellation Virgo is overhead at midnight in March or April, whereas Pisces rises with the sun during this time of the year. Both Virgo and Nisan are the first in their respective orders. Recall that this first constellation group presented the idea of a son who would be borne by a virgin. He would be a shepherd but would also go to war against an enemy.

In this conflict he would appear to lose the battle, ending in His death on a cross. As we take a closer look, however, we find that the Son is seen interacting in battle with Victima, another decan. In this battle He is the victor. But here in our first constellation group this victory is not so apparent. It is only revealed to those who are looking closely.

The feast of Passover is a time when God's people should be looking for signs of their redemption, yet not very many noticed what happened two thousand years ago when the prophecy of this feast was fulfilled. Yeshua came and died on a cross on this very feast, just as the Virgo's battle and the cross in the sky foretold.

Two Feasts, Close Together

During this same month two other feasts occur. The Feast of Unleavened Bread is celebrated on the 15th day of Nisan, and Firstfruits takes place within the week. The primary theme of these feasts teaches us that the One who will become the Passover Lamb will also serve up His life on the cross. Through His death He will then reestablish the authority to take away the sins of the world and promise life after death for those who believe in His work, trust Him, and rely on Him for their deliverance.

Pisces, featuring a captured and bound fish and Andromeda in chains, marry perfectly with this theme. This constellation reveals that — without divine intervention — mankind is bound and sentenced to death. Recall that Cetus (both Cetus and Aries interact with this constellation group), a decan lying underneath the captured fish is attempting to carry away the prize. But a lamb (Aries) intervenes on behalf of the fish, which represents mankind. This is also exactly what these two feasts suggest. The sacrifice of the Lamb will save mankind from their sins.

But what about the life after death promise of Firstfruits? Another decan interacting with this group is Perseus. This warrior provides shoulders on which the woman is standing. Via her broken chains, this image prophesies that through a great battle her bonds have been cut and she has been set free by the strength of another.

Of course, this warrior decan represents our Savior whose name, Yeshua, means the same. We have been released from our sentence of death and Firstfruits offers us the hope of walking in the footsteps of our Redeemer. This of course includes resurrection from death.

The last decan, Cepheus, pictures a king sitting upon his throne and reveals what the Bible prophesies concerning God's own Son.

> The Lord says to my Lord: "Sit at My right hand until I make Your enemies a footstool for Your feet." [2] The Lord will stretch forth Your strong scepter from Zion, saying, "Rule in the midst of Your enemies."
>
> —Psalm 110:1, 2

> And He who sits on the throne said, "Behold, I am making all things new." And He said, "Write, for these words are faithful and true."
>
> —Revelation 21:5

By His work on earth He is able to put His enemies under His feet, and He will make all things new, including the everlasting life given to every believer.

Celebrating the Work of God

The fourth feast is Shavuot. Recall that it is celebrated 50 days after Firstfruits, the third feast, which would place its celebration in the Hebrew month of Sivan. This occurs in the last few weeks of spring, usually in the early days of June or at the very end of May. June's overhead constellation group, at midnight, is Scorpio. That same month, Taurus rises with the sun.

As we have learned, this feast is the celebration of the work of God. We can take refuge in the knowledge that He has won the battle against

Satan. As we please Him with our conformity to His instructions, our destiny is with Him. Shavuot is seen in the pictures of these constellation groups. Taurus and Ophiuchus are doing battle against our enemy and are preventing Satan from reaching his goal. The right to rule the creation is represented by the crown over Serpent's head. Our destiny is the lap of the Shepherd seen in Auriga.

Satan's destiny is the decan Eridanus, the river of fire. During this time of the year, someday in the future, biblical prophecy suggests that the Antichrist or the beast, personified by Satan, will make his last move. He will reach for the crown that will give him the right to rule and reign over the whole earth. We know that he will take the very throne of God, which will be placed in the newly constructed temple in Jerusalem. He will then rule for three-and-one-half years, which means that Satan's kingdom will end sometime during the fall feast, probably in the month of September or October. Is it accidental that overhead during the feast of Shavuot is a scene of a huge snake reaching for a crown, symbolizing the same authority to reign?

Aquarius and Leo are the constellation groups during the September/October period, the time of the fall feasts of Trumpets, Yom Kippur, and Sukkot. Aquarius is overhead and Leo rises with the sun during the fall festivals.

Many biblical scholars, including Jewish rabbis, believe that the Bible foretells that the Messiah's coming will occur during the Feast of Trumpets, when Aquarius is the dominate voice seen at night. Recall that this group of decans prophesied the coming of the Messiah, as pictured in Pegasus.

The white horse Pegasus carries God to earth. He is the one described in Revelation 19, with an army behind Him and a two-edged sword protruding from His mouth. This represents how God will destroy His enemies; i.e., with His voice. He will bring all things to an end just as it all began. He created the heavens and the earth and all the life thereon with the word of His mouth. He will remove all those who now oppose Him in the same way.

God Will Come Back

These three feasts speak a very clear message. God will come back for those who love Him, and He will deliver judgment to those who oppose Him. Yom Kippur suggests the idea of finality, cleansing, and marriage. Sukkot, the last of the seven feasts, conveys the idea of completion and purity, of a place now restored back to the Garden of Eden's perfection. This will be a time of peace in which God and mankind can once again dwell together, with mankind restored to right standing with his Creator.

Aquarius is the provider of the living water from His urn. Mankind, represented by the fish who is gulping up the urn's contents, thus finds renewed life and restoration from sin. The stars and His Word confirm this idea even by the timing of the feasts and the appearance of the constellation's messages. Everything in God's kingdom represents order and coordination. Everything and everyone in His kingdom works together and basically complements everything else.

Leo, appearing as the sun arises, continues to build on this same theme. This lion is the Messiah who will come and deal with Hydra by pouring out judgments on Satan's kingdom, and by destroying our adversary's army. Corvus and Crater, decans within the Leo constellation group, represent the coming judgment upon Satan, his army, and his kingdom. The Feast of Trumpets is a call to war — this very war between God and Satan. The stars suggest that when the authority and rule of Satan is finally done away with a time of peace will reign supreme once again.

The beginning and ending of this circle of constellations, represented by Virgo and Leo, also suggest order and coordination. Virgo begins the story that includes Virgo and Leo working together to bring forth the Son, our Messiah. But it was obvious that she needs some help, for her state of being was not compatible with God. Virgo was fallen and in need of restoration.

Leo, the last constellation, is now at her side after a long journey of battle, death, renewal, and judgment. Now our Lord also stands by his/her side. This coming together is the perfect picture of a Hebrew marriage. In ancient times in Israel, the bride and groom would come together and make a contract called a *ketubah*. It would include obligations for both the bride and the groom to accomplish to prepare themselves for the wedding and marriage. Both would look forward to the time when they would no longer be separated but be united, working side-by-side in unison through marriage.

Sukkot, the Bride and the Groom Together Forever

Our concluding scene is exactly that. Our Groom and Virgo, His bride at the consummation of His plan, are at each other's side. This is the place God has intended for us to be all along. Only by rebellion did we take a detour. But through His commitment to win back the right to save us from ourselves, we now find our chance to once again be at His side in the best relationship offered in the universe. To not know and love the Creator of everything — oh what a missed opportunity that would be!

I don't believe it's accidental that Passover, the first feast in the first month of the Hebrew calendar, is represented by Virgo appearing overhead at midnight. Likewise the last feast, Sukkot, occurring in the seventh or eighth month, is represented by Leo rising with the sun at that time.

In both of these constellation groups God is the primary focus. In the first He appears as the humble Son who will offer Himself on a cross. In the second our Messiah comes as the conquering King. The voices in the sky proclaim that He will be the first to come to our rescue, but also that the King shall come in the end and lift us up by defeating Satan. These prophecies echo His exact words in the Bible as well.

> "I am the Alpha and the Omega," says the Lord God, "who is and who was and who is to come, the Almighty."
>
> —Revelation 1:8

Sukkot is the last of the godly celebrations each year. It was a time when the Hebrews would build sukkot (small shelters) in their fields and live in them for seven days. The feast was meant to remind them of their ancestors' trek through the wilderness, during which they lived in tents. The important implication was that this was the time when God came and dwelt with His people in the middle of their camp, in the tabernacle.

This very feast is also a prophecy foretelling the final parts of God's plan. His end game is to once again be at our side, dwelling on earth and ruling His creation with us.

Chapter Fourteen

Historical Events

In God's amazing wisdom He is able to speak to us in many ways. He certainly has given us His Word, which gives us great detail in how to please Him. But we have also learned that He has not turned a blind eye to those who have not been fortunate enough to have access to the Bible, nor to those who lived in ancient times before His Word was given to the world.

Elohim has revealed Himself to all. Psalm 19 proclaims that there is nowhere the messages of the heavens are not heard. The voices of the heavens go out to all the earth, telling of the splendor and majesty of its Creator. Thus Paul warned in Romans 1 that none will have an excuse because God's eternal power and holy nature are clearly revealed through what has been made. We should heed the messages He has provided, for they all agree. There is no conflict between the messages of the heavens and the messages from God's Word.

On a more personal level, God has provided words of encouragement in all of our lives, especially love or guidance at just the right moment. Thank God that His instructions are so readily at hand. We do not have to depend on our feelings or use subjective emotions to gain heavenly guidance. His words have been written and fixed and are not subject to our whims, if we use proper means to understand them.

But what about those less fortunate, those who have lived before the Bible had spread throughout the world? Could the creation guide us in the same manner as His written directives? I think so. As most can agree, many times God has used the beauty of His creation to speak to us, to inspire our spirits, or to encourage us to do some great thing.

What about the godly voices that come from the heavens? Have they provided these same words of guidance at just the right time? If we look back in time to various moments when people have needed surety that they were not alone, that God was right there by their side, did God provide a message from above for them as well?

What About Esther?

Let's explore the lives of a few familiar biblical characters to see whether God saw fit to talk with them even if they didn't have access to a Tanakh. For example, Esther is a familiar figure in the Old Testament. Recall that, in the book named after her, she was chosen to be Ahasuerus's queen. At the same time, the evil Haman, who was given great authority in the kingdom, hated the Jews. Through his high position he was able to enact a law that would have exterminated all the Jews in the land.

Meanwhile, Esther, who had previously kept her Jewish heritage a secret, was asked to confront the king on behalf of her people, knowing that to approach the king unrequested was often a death sentence. Yet with God at the helm, all went well and the evil intentions of this law were thwarted. And God's people were saved.

Today the Jews celebrate Purim in remembrance of the deliverance that God provided for them. But what about Esther? Was God supporting and encouraging her, telling her that He did, in fact, have everything under control? Her crisis came to a climax in the month of March. It is easy to imagine Esther, during the evenings leading up to her confrontation with the king, looking up to heaven and asking for help and encouragement from God above.

As she looked up she would have seen Leo, the dominant constellation during that time of year. What could the lion tell her that would be a timely message for this frightened queen, who had the weight of the world on her back? Recall that this constellation, with its decans, is a message of hope. And that our enemy, whether it is fear, depression, loneliness, or opposing power and authority, such as Haman, is in the hands of God. God is in control and His plans will win out in the end.

Esther could have looked up and observed the lion with its front paws on the head of Hydra, representing Satan. To her , Hydra was Haman. She could have been encouraged by seeing the messages from the other decans as well. Corvus and Crater, the bird eating the Hydra and the bowl pouring out judgment on the back of her enemy, could have given Esther a clear picture of the future and perhaps just enough encouragement to confront Haman in the court of her king.

What happened? Satan's demonic schemes were turned back on Haman's head, saving the Jews. He was hung on the gallows just as the heavens proclaimed. As Leo brought intervention and salvation, he also brought judgment to the enemy. God did not leave Esther without guidance and encouragement from above.

What About Moses?

Moses was chosen by God to lead His people out of the land of Egypt, where they had been enslaved by Pharaoh. The beginning of his trial occurred during the same month as Esther's ordeal, although his happened about a thousand years earlier than hers.

Moses was definitely a man in need of encouragement. Just as Esther did, he could have looked up and heard basically the same messages from God — messages that would have given him the strength to do what God was calling him to do. Instead, when given instructions by God at the burning bush, he tried to back out of the deal by reminding God that he was not a good orator. This lack of confidence revealed his need of support from just about anywhere, even though God was backing up His face-to-face message with what He'd already written in the sky.

The final confrontation between Moses and Pharaoh, and the ultimate release of God's people from their enslavement, occurred in the next month, the Hebrew month of Nisan. This month usually occurs in our April time frame. Of course, Virgo is overhead, and Israel would have just received God's instructions to celebrate the first Passover.

The message embedded within this annual convocation was that a Savior would come and substitute Himself as a sacrifice for mankind's sins. His death would provide a covering for them. Symbolized by the Passover lamb, with its blood painted on the doorposts of their homes, the Messiah's blood did save the Israelites from the judgment that fell upon Egypt's firstborn that very night.

Virgo complements this message, revealing the Savior, the Lamb, as the Son who would come via a virgin birth. He would have a double nature and be both man and God, and He would die on a cross to provide the needed covering. Using the Passover lambs' blood, the enslaved Israelites painted an ancient tav on each of their doorposts, in the shape of the last letter of their aleph-bet. In doing so they might have recalled the cross in the sky just below Centaurus. The ancient letter tav, drawn in the form of a cross, meant "the sign of the covenant."

This covenant was not a new concept to the Hebrews. Their parents had taught them that God had made a blood covenant with their people long ago, through their patriarch, Abraham. The events of these days, plus all the messages and promises from God, both ancient and new, could have coalesced in their minds and completed a fuller picture of the plans of God that He would work out through them. If only they would take the journey.

What About Joshua?

Joshua, crossing the Jordan and invading the Promised Land, quickly attacked Jericho. This city fell during the same time of the year that the Israelites left Egypt some forty years earlier. The same message of hope and salvation further supported them in their efforts to fulfill God's directive in confronting the giants in the new land. Thus the message of the centaur would have complemented the words given to Joshua: "I will go before you and help you fight your battles."

Recall that the decan Centaurus, in the constellation Virgo, complements Virgo's theme. That theme is that God will send a Son who will provide hope and, in the end, victory over sin and death. The Israelites who entered into the Promised Land in the face of the giants and the well-trained armies that dwelt in the land could have looked up as they entered and been encouraged by the constellations that were overhead at night, bringing them a promise of a victory.

What About Hanukkah?

The Jews have celebrated another feast for more than two thousand years, remembering an ancient enemy who opposed their God and His people. This pagan king, known as Antiochus Epiphanes, made the temple impure by slaughtering pigs on its holy altar. He also persecuted the Hebrews by annulling their laws and killing all who disagreed. In fact this king, the leader of the Seleucid Empire to the north of Israel, is an archetype of the Antichrist, or False Messiah, known as the beast in the book of Revelation as personified by Satan.

The Maccabees, a group of Jews who rose up to oppose this evil ruler, battled against him, overthrew his army, took back the temple, and cleansed it from its impurities. Hanukkah, celebrated in the month of December, commemorates the righteous acts of the Maccabees and the miraculous events brought about by the hand of God in support of His faithful. After winning back the temple they needed to purify the temple grounds and produce fresh oil for the menorah, using a special process that took eight days. The miracle was that they had sufficient oil for keeping the menorah lit for only one day, but through God's intervention the oil lasted for the entire eight days — long enough for them to produce a new batch. The menorah stood in the inner court of the temple of God and was recognized as a sacred light that represented the light of God.

Also, history suggests that the army of the Jews was not an army at all. Rather it was a ragtag collection of people willing to trust God in the face of the most powerful army of the day. Just like Israel in 1948, it had no chance of winning except for the ally they hoped would show up. That was God Himself, of course.

If they had looked up during those fateful days, the night sky would have informed them that they had already won. Their ally had written them a message centuries earlier encouraging them to go ahead, for God would, indeed, show up and fight for them. Taurus the bull, whom we now know is another representation of God, will fight for His people. The decans tell of His victory against an evil lion, Satan, who prowls around looking for whomever he can devour. Auriga also comforts those in battle by picturing the kids of a goat in a Shepherd's lap.

Auriga, One of the Decans of Taurus

All of this simply proves that, even though we may be called to battle, we can actually rest in His lap while serving Him. The Maccabees, and Israel today, could have entered into this rest in the face of their monsters as well. Eridanus would have informed these ancient Israelites that God will not stop at just defeating their enemies. He has a special place reserved just for His opponents — the Lake of Fire via Eridanus, one of the decans of Taurus and the river of fire itself.

The fall of Jerusalem in 586 BC occurred in the month of Av on the Hebrew calendar. In fact, many of the darkest events in Jewish history have occurred during the first nine days of this month. Tisha B'Av, the ninth of Av, commemorates these events, including the following:

1 Av (1273 BC) – Death of Aaron
The High Priest Aaron, who was the brother of Moses and Miriam, died at age 123 of the Hebrew year 2488 (1273 BC). This is the only yahrzeit (date of death) explicitly mentioned in the Torah (Numbers 33:38).

5 Av (AD 1572) – Death of Rabbi Isaac Luria
Rabbi Isaac Luria Ashkenazi, known as "Ari HaKadosh" ("The Holy Lion"), died on the 5th of Av of the Hebrew year 5332 (AD 1572). He was born in Jerusalem in 1534, and spent

many years in secluded study near Cairo, Egypt. In 1570 he settled in Sefad, where he lived for two years until he died at age 38. During that brief time, Luria would change the study of kabbalah. He has since come to be considered one of the most important figures of Jewish mysticism.

7 Av (586 BC) – First Temple Invaded
After a month of fighting in Jerusalem, King Nebuchadnezzar's armies of Babylonia broke through, into the holy temple, where they feasted and vandalized until the afternoon of Av 9, when they set the temple on fire.

7 Av (AD 67) – Civil War in Jerusalem
Inside the besieged city of Jerusalem, fighting broke out between different Jewish factions divided on the question of whether to fight the Roman armies who had encircled the city. One group set fire to the city's food stores, which is said to have quickened starvation. Jerusalem would fall three years later.

9 Av (circa 1446 BC) – Rebellion of the Israelites in the Wilderness
Moses sent twelve scouts to spy out the land of Canaan. Ten of the scouts gave negative reports, causing despair and lack of trust among the Israelites. For this, God decreed that none of that generation would enter the Promised Land except Joshua and Caleb, the only two scouts who believed God would enable them to take possession of the land.

9 Av (586 BC and AD 70) – Holy Temples Destroyed
The first and second holy temples, which stood in Jerusalem, were both destroyed on Av 9. The first temple was destroyed by the Babylonians led by Nebuchadnezzar, and the second temple by the Romans led by Titus. The temples' destruction represented the beginning of *galut* (exile).

9 Av (AD 71) Romans Replace Jerusalem with Aelia Capitolina
Turnus Rufus plows the site of the temple. Romans build pagan city of Aelia Capitolina on the site of Jerusalem.

9 Av (AD 133) – Failure of Bar Kochba's Revolt and the Fall of Betar
Betar, the last stronghold of the Bar Kochba rebellion, fell to the Romans on the 9th of Av after a three-year siege. 580,000 Jews are said to have died either by starvation or the sword, including Bar Kochba, the leader of the rebellion.

9 Av (AD 1096) – First Crusade
The First Crusade officially commenced on August 15, 1096 (the 9th of Av), killing 10,000 Jews in the first month and destroying many Jewish communities. The crusades ultimately took the lives of 1.2 million Jews.

9 Av (AD 1290) – Jews Expelled from England
Jews were expelled from England by the Edict of Expulsion of King Edward I. Their property and possessions were confiscated and pogroms ensued. They were not legally permitted to return for 350 years.

9 Av (1306) – Jews Expelled from France

9 Av (1492) – Jews Expelled from Spain
The Spanish Inquisition culminated with the expulsion of the Jews from Spain by King Ferdinand and Queen Isabella, destroying many centuries of Jewish life there. Thousands of deaths resulted.

9 Av (1914) – World War I Began
WWI began on the 9th of Av when Germany declared war on Russia. German resentment from the war set the stage for the Holocaust.

9 Av (1941) – "The Final Solution"
SS Commander Heinrich Himmler received formal approval from the Nazi Party for "The Final Solution," resulting in the deaths of 50% of the world's Jews.

9 Av (1942) – Warsaw Ghetto Deportation
On the eve of Av 9 the mass deportation of Jews to Treblinka began.

9 Av (1994) – AMIA Bombing
Arabs bombed the Jewish community center (AMIA) in Buenos Aires, Argentina, which killed 86 and wounded more than 300. This was the deadliest attack in the Jewish Diaspora since the Holocaust.

9 Av (2005) — Forced Evacuation of Gaza's Gush Katif Communities
Thousands of Israelis were evacuated from 25 towns and settlements in the Gaza Strip in the summer of 2005 as the beginning of "dividing Israel for peace," Israel's disengagement plan.

10 Av (AD 70) – Holy Temple Burns
The Romans set the temple on fire on the afternoon of Av 9. The mourning practices for Jews of the "Nine Days" are observed through the morning hours of Av 10.

12 Av (1263) – Nachmanides' Disputation
King James I of Aragon (Spain) ordered Nachmanides (Rabbi Moses ben Nachman, 1194–1270) to participate in a public debate, held in the king's presence, against the Jewish convert to Christianity, Pablo Christiani. His defense of Judaism and refutations of Christianity's claims served as the basis of future disputations through the generations. Because his victory was an insult to the king's religion, Nachmanides was forced to flee Spain and came to Jerusalem.

15 Av (AD 148) – Betar Dead Buried
The Betar fortress was the last holdout of the Bar Kochba rebellion. Betar fell on the 9th of Av, Hebrew year, 3893 (AD 133). Bar Kochba and many thousands of Jews were killed by the Romans. The Romans would not allow the Jews to bury their dead for 15 years afterwards. The dead of Betar were brought to burial on Av 15 of the year Hebrew year 3908 (AD 148).

An additional blessing (*HaTov VehaMeitiv*) was added to the "Grace After Meals" in their commemoration.

17 Av (1929) – Hebron Massacre
Sixty-seven Jewish adults and children were killed, and others were wounded, raped, or injured by Arabs in Hebron. The Arabs rioted for three days, yelling out cries to "Slaughter the Jews." The survivors fled to Jerusalem. This massacre destroyed the ancient Jewish community of Hebron, which had existed there, relatively peacefully, for centuries. Jews would not return until after Israel's capture of Hebron in the 1967 Six-Day War.

21 Av (1918) – Death of Rabbi Chaim Brisker
Rabbi Chaim Soloveichik of Brisk (1853-1918) died on this day. He was a famous Talmudic scholar and Jewish leader.

Capricorn and its decans portray the idea of sadness. This constellation appears overhead in the month of Av. Aquila, one of the decans of Capricorn, is the dying eagle falling out of the sky, representing God who is no longer overhead watching over us. The temple and its two destructions in this month communicate the same message. By destroying the dwelling place of God Satan took satisfaction in knowing that his opponent no longer resided with man.

Sagitta, another Capricorn decan, possibly representing the arrow from the bow of the evil and deceptive rider of the white horse in Revelation 6, thought that his arrow might have done its dirty work. With God falling and out of the way, Satan could get on with his plan to conquer and rule the earth. Of course, Satan is wrong but he still has his delusions, though his time is short.

I always hold my breath during this month, hoping that nothing untoward happens to the people of Israel during its days. But I am also reminded, by the voices from above, that it is through times of hardship that new life, maturity, and renewed relationships are also born. Capricorn, itself a union of man and God, offers the hope of restored friendship with our Creator by its two-part character — part goat and part fish. Meanwhile, Delphinus, the dolphin jumping from a wavy surface of water and the last decan of Capricorn, reminds us of the life that suffering sometimes produces, especially suffering at the hand of God.

Conclusion . . .

These twelve constellation groups all speak of the subjugation of all things to the Messiah the King. They all bring messages telling that everything that opposes Him will surely be brought low. Genesis 37 reports to us a dream that Joseph had. He was one of the twelve sons of Israel. This prophecy told of the stars, moon, and sun all bowing down to him. Upon his retelling of the dream to his family, you can imagine their reaction. It was not very good.

But in Scripture, Joseph's life is presented as an archetype of Messiah Yeshua. As Joseph brought salvation to his family, the tribes of Israel, so will Yeshua bring salvation to His family, Israel. The interesting point here is the image that appears in Joseph's dream. Even the heavenly bodies are in submission to our Savior.

And they also tell of this same salvation that will come.

Chapter Fifteen

Putting It All Together

If you have read my other books you are well aware of the concept of the menorah. Its set of seven lights lit the inner court of the tabernacle as well as the temple. The fourth light was known as the shamash and was the middle light, with three other lights on each side. In Hebrew thinking it represented the Messiah, because all the other lights were lit by its fire. In Hebrew the sun is called the shamash and metaphorically represents God as well.

This concept of seven is interspersed throughout the Hebrew text of the Bible, with the idea that the number seven ushers in completion and perfection, or wholeness. The creation week certainly is an example of this. On the seventh day God rested because He had completed His work. It was whole and perfect, ready for man to live in and thrive.

On the fourth day, the focus of His work centered on our sun, the shamash, along with the rest of the lights in the sky. This is not accidental. God is reminding us that He should be the center of our lives as we revolve around Him. In Him we procure warmth, light, and direction each day. This was part of God's plan from the beginning and will continue to be.

As we are seeing, there are way too many connections, references, and concepts in the Old and New Testament that are intertwined. There are not two plans, nor are there two ways or two peoples that God will call His bride. A person is either a believer and a Jew or a believer and a Gentile. As we have already explained, in God's Word, one is called an olive tree with roots, a trunk, and branches, and the other is grafted into this same tree with the same ultimate identity.

Some are skeptical of the idea that all believers, no matter their heritage, are represented by the word "Israel." One of the purposes of this book is to shed some light on the fact that, no matter what we might have been taught, this idea has a solid biblical basis. Many passages support this principle, including this one in Exodus:

> Seven days there shall be no leaven found in your houses; for whoever eats what is leavened, that person shall be cut off from the congregation of Israel, whether he is an alien or a native of the land.
>
> —Exodus 12:19

Here God is revealing that the congregation is composed of both strangers and those born in the land. The people who joined with the descendants of Israel and left Egypt behind did not separate themselves from the Hebrews by forming a separate tribe (Exodus 12:38). Only twelve groups are listed. From time to time in Torah, non-Hebraic people are mentioned and are always invited to identify with Israel. The same laws God gave to the Hebrews were also given to these foreigners. Evidently these people integrated into the twelve tribes by acquiring the identifying markers that God uses to identify His people.

> "How blessed is the man who does this, and the son of man who takes hold of it; who keeps from profaning the Sabbath, and keeps his hand from doing any evil." [3] Let not the foreigner who has joined himself to the LORD say, "The LORD will surely separate me from His people." Nor let the eunuch say, "Behold, I am a dry tree." [4] For thus says the LORD, "To the eunuchs who keep My Sabbaths, and choose what pleases Me, and hold fast My covenant, [5] to them I will give in My house

> and within My walls a memorial, and a name better than that of sons and daughters; I will give them an everlasting name which will not be cut off. [6] Also the foreigners who join themselves to the Lord, to minister to Him, and to love the name of the Lord, to be His servants, every one who keeps from profaning the sabbath and holds fast My covenant; [7] even those I will bring to My holy mountain and make them joyful in My house of prayer. Their burnt offerings and their sacrifices will be acceptable on My altar; for My house will be called a house of prayer for all the peoples." [8] The Lord God, who gathers the dispersed of Israel, declares, "Yet others I will gather to them, to those already gathered."
>
> —Isaiah 56:2–8

Honoring and observing God's Sabbaths is one of the hallmarks of this obvious end-times passage. With respect to the Law and Passover, Exodus 12 gives us some additional insight.

> All the congregation of Israel shall keep it. [48] And when a stranger shall sojourn with thee, and will keep the Passover to the LORD, let all his males be circumcised, and then let him come near and keep it; and he shall be as one that is born in the land: for no uncircumcised person shall eat thereof. [49] One law shall be to him that is homeborn, and unto the stranger that sojourneth among you.
>
> —Exodus 12:47–49, KVJ

Along with others, these passages make it clear that non-Hebrews can become Israel by observing the same ordinances that God gave to His people. Paul makes the point that there are indeed, two definitions of Israel.

> But it is not as though the word of God has failed. For they are not all Israel who are descended from Israel.
>
> —Romans 9:6

He seems to be implying that the Jews referred to themselves as Israel. But Paul proposed that not all of Israel's descendants were in fact

Israel, for this was the Israel that God intended His people to be part of through faith. Many already have, and many more will become part of Israel in the future. But Paul is making it clear that faith is the means by which others can be identified with Israel. And, that through faith, salvation has come to the Jew first but also to the rest of humanity.

The congregation of Israel should be a happy group. No matter what their heritage, they have all been invited to share the blessings of the plan of God. And it is unfolding before our very eyes. In fact, the very endgame and completion of His plan is upon us. We should be making our final preparations. Yes?

What's Up With the Twelve Sons?

Jacob/Israel had twelve sons. God could have given him just three, or seven, or fifteen for that matter. But He gave him twelve. Was this an accident or was there a purpose behind it? Could God be drawing our attention to the stars and the twelve constellation groups, using the fact that Israel had twelve sons?

We have learned that God used the twelve signs in the sky to describe His plans for His creation, with special emphasis on mankind. These signs develop, bring to conclusion, and promise a perfectly restored order. Ultimately, they suggest that a righteous King will once again reign over the universe.

Twelve sons were given to Israel. They were also given the responsibility to restore righteousness. To help them accomplish this they were given access to God, who gave them His principles for holy living. By asking them to integrate His ways into their lives God could then affect all the nations, thereby establishing holiness throughout the earth.

The number twelve has some interesting biblical implications and usages. Twelve is a perfect number, signifying perfection of government. It is found as a multiple in all that has to do with "ruling." The sun which "rules" the day, and the moon and stars that "govern" the night, do so by their passage through the twelve signs of the Zodiac

(Mazzarot in the Bible), which completes the great circle of the heavens of 360 degrees or divisions (12 x 30), and thus govern the year.[28]

God also called twelve apostles. He established twelve foundations, twelve gates, and twelve pearls in the heavenly Jerusalem and twelve gems on the High Priest's breastplate. In the book of Revelation He sealed 144,000 Israelite males, all virgins. This number is derived by multiplying twelve times itself.

What is God saying by using the number twelve in describing His plan of restoration, meanwhile grouping His people into twelve sections? His bride will be given the right to rule over His creation once again, along with Himself. But this time the restoration of His bride will also bring perfect, righteous governance back into the earth instead of chaos. No wonder God used twelve groups of constellations, for they foretell this same union of God and mankind in a restored rule.

The Blessing of Authority

The Hebrew word for bride is *kallah* and is spelled *kaf, lamed, hey*. The pictographs of these letters say "behold the blessing of authority." This is exactly what the governance of the bride of Yeshua is supposed to bring forth.

Kallah means "completion and perfection, to be prepared, made ready" but also means "consumption and destruction, to be destroyed, to perish and to waste away." What's going on here? Why these two opposing meanings?

The book of Revelation reveals two brides. The first is God's bride who will rule and reign with Him. She will be made complete and perfect, ready for her Groom. On the other hand, Satan's bride is fit only for destruction. She has allied herself with the destroyer and so she will waste away, perishing along with her groom in the place of destruction.

The Unity of the Bible

The biblical representation of God's bride as "twelve" in the context of the twelve sons of Israel points toward a unique uniformity. This same unity also includes the messages contained within the ancient pictures in the sky.

This unity is exactly how the Bible is put together as well. In reality, there is neither an Old nor a New Testament, but a collection of many books that are tied together by culture, heritage, and principle. All find their roots in the same God who presented all of this truth to mankind.

What separates believers from unbelievers, other than their trust and reliance on Yeshua's blood? It is this very plan we have discovered in the Bible, as well as in the stars. This God-created plan includes a choice for us. That choice offers two different relationships. The first choice is with God, the Lion who wins the battle. The second choice results in another relationship as offered by Satan, the dragon. He is the one who loses at the hand of the Lion.

As exposed in the stars, the battles between God and Satan (the Lion and the dragon) started in Eden. At the center of this very special garden grew a tree with a unique name: the Tree of the Knowledge of Good and Evil. This was the tree that God told Adam not to eat from.

The word for knowledge in Hebrew is from the root *yada*. This word means knowledge yet is a very special word. Along with the idea of information, wisdom, and knowledge it also conveys an intimate tone. It is the word used in the original Hebrew when the biblical text says that Adam knew Eve and begat children. This word conveys the idea of intimate knowledge, not just a list of facts.

The question then would be, why would God make knowledge sin? Why not name this tree something else, like Apple, or Pear, or Tree of Voices? Why did He choose the name He did? Is he trying to draw our attention to something important? Because He chose that name is God trying to say that He is against knowledge?

Adam and Eve's Mistake

The above questions remind me of an experience I once had. I had asked for prayer after the service one morning, when I was told by the one praying for me that I needed to get my mind out of the way for God to do His work in me. What she was proposing was that I was thinking too much. I needed to just blank out my mind and let God have control. At that point I realized I had come to the wrong place for prayer. At the time I was asking God for more of His spirit in my life. I desired a freer flow of the gifts from God as spoken about in 1 Corinthians 12. What this person was proposing was for me to repeat the mistake that Adam and Eve made.

Satan offered knowledge to his admirers as well, in an attempt to short circuit God's plans. But God took a different approach — He wanted Adam and Eve to learn *with* Him. He wanted to teach them His truth as they walked together in the garden. Because, only through relationships with Him are the power and insight and purposes of God's plans revealed.

Satan cut through this slow but rich process and ruined God's fun. Have you ever been telling a joke or story with a cool ending, only to have someone else jump in and steal the punch line? This is what Satan did, but much worse. He proposed to Eve that she could bypass the relational development with God that came with the truth, and then he substituted himself for God.

In other words, knowledge from God brings insight and wisdom, and it allows us to please Him by fulfilling His will for our lives. Without God, knowledge leads to bondage and sin, and intimacy with evil. The first and second chapters of Romans speak to us in this regard. Paul tells us that people willingly chose to ignore truth and believe a lie (Romans 1:25).

Now . . . you may ask how someone could be so stupid on purpose. Knowledge without God leads you away from God and sets you in the lap of pride and self-aggrandizement. Knowledge learned at the hand of God leads to wisdom, insight, and intimacy with Him.

From the moment of the first deception in the Garden of Eden there has been a battle between two opposing bodies of knowledge. One originates with God while the other originates with Satan. The second one is a perversion of God's truth, and was given to mankind to pull him away from the real truth and ultimately to bring about his demise. The entire Bible is a history of battle between these two bodies of knowledge.

Two Basic Choices

Remember the twelve spies that Moses sent into the Promised Land? Why did two of them see things differently than the other ten? Both groups observed the same thing: the Nephilim, who were giants, but also the bounty of the land. The two spies who brought back a good report perceived God's plan. They perceived His purposes. They saw the blessings and the power and the victory that would come from God if the Israelites would simply accept what He was offering.

The others were blinded by their personal fears. Instead of being guided by their trust in God, they were deceived, discouraged, and misdirected by their own bondages. They missed out on a new inheritance, a new life, and the opportunity to be part of an ancient plan God had custom-made just for them. Instead, because of their lack of trust and reliance on God they spent the balance of their lives wasting away in the wilderness. Ultimately they died because of their lack of preparation.

We too have these same choices. We can confront the giants in our own lives, secure in the knowledge that God will be by our side helping us all the way. Our reward will then be the same as that of the spies who came back with the good report, a custom-made blessing fashioned just for us. Or, we too can fail. Trusting in God is hard but it will lead to the Promised Land.

As we live out our lives, the environments that we are raised in and the things that we are taught, followed by the events that we live through, all cause us to develop various world views. It is these individual perspectives that give us our uniqueness, but also our blinders and prejudices.

The reporter we met in the story in the introduction of this book reflected this same cause-and-effect. His perspective was certainly not a correct one, but he may have thought he was properly reporting the news. And most of us do the same thing. Most of us do not purposely lie. This is why, when an investigator interviews multiple eyewitnesses to a crime, many times they will contradict each other.

Sometimes these contradictions can be accounted for because the witnesses were not making their observations from the same vantage point. But as was true of our reporter, sometimes our prejudices get in the way and cause us to "see" what is actually a perversion of the truth. These perversions then make it more difficult for us to hear God's voice and observe His instructions, which would guide us and bring us blessings and safety.

Two Basic Truths

Our goal in this book has been to show that God has delivered to all of us a plan for the destiny of mankind. We have learned that this plan can be found not just in the Old Testament but also in the New. In fact, these two testaments work together, like chapters in a book, completing the telling and foretelling of this ancient story.

However, God has not given us His plan in written form only. He has also revealed the same story in the sky for all to see. Included in His plan is a unique role carved out for each one of us. That role fully complements the skills and abilities He gave us. We are more than just pawns who — despite our personal uniqueness, rights, and wills — get swept up in plots and strategies beyond our control. It has long been God's desire to include us as His partners in helping bring His goals to fruition.

It is ironic that the all-powerful God asks us to play an important role in creating and shaping His plan for our life. Why does He need us? This irony reveals His very heart. We are important beings of much value to Him. Without you there will be a hole in His creation, incompleteness in His army, a gap in the very body of His bride.

Sadly, we have enemies who are trying to prevent this partnership from forming in the first place. And if that doesn't work they then try to turn us into ineffective hypocrites without any understanding, without knowledge or truth. Thus they bring confusion into the equation, which prevents us from making clear observations and decisions.

A majority of believers see this opponent as Satan. But most of the time, what prevents us from correctly understanding reality is birthed from within. That is why God goes out of His way to urge us to be seekers after truth. That is why He gave us the Holy Spirit, whose goal is to guide us into all truth (John 16:13; 1 Cor. 2:13). And truth has this very interesting quality built in. It repels darkness and deception and allows us to see even more truth.

Throughout the Bible we find an unfolding revelation and presentation of two "truths." Each one is vying for your attention and investment. Both Satan and God offer their plans. Have you ever wondered why God chose in the beginning to make eating from the Tree of the Knowledge of Good and Evil the only sin? Why not have "sin" result from touching those rocks over there, or some other action even more random? He chose this tree, having fruit representing the two "truths" – one good and one evil.

We have been given a choice between two truths. By choosing God's plan and His truth, which leads you directly to Him, you also chose Him. Yahweh is not into arranged marriages.

Endnotes

[1] John Klein, Adam Spears and Michael Christopher, *Lost in Translation Series, Volume 1: Rediscovering the Hebrew Roots of Our Faith* (Bristol, TN: Selah Publishing Group, 2007), p. 30.
[2] H.W.F. Gesenius, *Gesenius' Hebrew-Chaldee Lexicon to the Old Testament* (Grand Rapids, MI: Baker Books, 1979 [originally published in 1847]), p. 36b.
[3] Gesenius, p. 142a.
[4] *Sefer HaChinuch–the Chumash/Stone Edition*, Rabbi Nosson Scherman and Rabbi Meir Zlotowitz, General Editors.
[5] Gesenius, p. 366a, b; 860a.
[6] Sir Robert Anderson, *The Coming Prince* (Lawton, OK: Great Plains Press, 1895/2012).
[7] Woody Allen, U.S. movie actor, comedian, and director.
[8] Gesenius, p. 457a, b.
[9] In this usage the word *ensign* refers to a flag flown (as on a ship) to establish nationality, often with a distinctive badge as part of the design. It also refers to a commissioned officer in the U.S. navy or coast guard, which is probably the more familiar usage in the modern era.
[10] Gesenius, p. 24b.
[11] http://www.idolphin.org/camp.html.
[12] Gesenius, p. 462a.
[13] http://en.wikipedia.org/wiki/Ecliptic.
[14] http://en.wikipedia.org/wiki/Axial_tilt.
[15]Kenneth Fleming, *God's Voice in the Stars* (Neptune, NJ: Loizeaux Brothers, Inc., 1981), p. 46.
[16] http://www.ldolphin.org/Nimrod.html.
[17] Gesenius, p. 833b.
[18] Gesenius, p. 367a.
[19] Danny Ben-Gigi, *Living Israeli Hebrew* (Scottsdale, AZ: Living Israeli Hebrew Publications, 2004), p.28.
[20] Fleming, p. 106.

[21] John Klein, Adam Spears and Michael Christopher, *Lost in Translation Series, Volume 2: The Book of Revelation Through Hebrew Eyes* (Bristol, TN: Selah Publishing Group, 2009), pp. 259, 260.
[22] Robert Scott Wadsworth and Daniel G. Stockemer, *A Voice Crying in the Heavens* (Salem, OR: Praise Publications, 1996).
[23] *Lost in Translation, Volume 1*, pp. 26, 27.
[24] http://en.wikipedia.org/wiki/Scarab_(artifact).
[25] Gesenius, p. 376a.
[26] Fleming, p. 136.
[27] John Klein, Adam Spears and Michael Christopher, *Lost in Translation Series, Volume 3: The Book of Revelation: Two Brides, Two Destinies* (Bristol, TN: Selah Publishing Group, 2012), p. 163–165.
[28] http://www.biblestudy.org/bibleref/meaning-of-numbers-in-bible/12.html.

Bibliography

Ben-Gigi, Danny. *Living Israeli Hebrew.* Scottsdale, AZ: Living Israeli Hebrew Publications, 2004.

Bullinger, E. W. *Witness of the Stars.* Grand Rapids, MI: Kregel Publications, 1893, 1967.

Fleming, Kenneth C. *God's Voice in the Stars.* Neptune, NJ: Loizaeaux Brothers, 1981.

Rolleston, Frances. *Mazzarot.* Whitefish, MT: Kessinger Publishing.

Seiss, Joseph A. *The Gospel in the Stars.* Grand Rapids, MI: Kregel Publication, 1979.

Wadsworth, Robert Scott and Stockemer, Daniel G. *A Voice Crying in the Heavens.* Salem, OR: Praise Publication, 1996.

Warner, Tim. *Mystery of the Mazzaroth.* Oakfield, ME: The Wild Olive Press, 2010.

www.ingramcontent.com/pod-product-compliance
Lightning Source LLC
LaVergne TN
LVHW050628100826
845148LV00011B/1776

9781589302914